Scale Formation in Heat Exchangers

Mohammad Varnaseri ·
Seyed Mohsen Peyghambarzadeh

Scale Formation in Heat Exchangers

Springer

Mohammad Varnaseri
School of Chemical Engineering
Amirkabir University of Technology
(Tehran Polytechnic), Mahshahr Campus
Mahshahr, Iran

Karun Petrochemical CO., Site 2
Petrochemical Special Zone
Bandar Imam Khomeini, Iran

Seyed Mohsen Peyghambarzadeh
Department of Chemical Engineering,
Mahshahr Branch
Islamic Azad University
Mahshahr, Iran

ISBN 978-3-031-52703-6 ISBN 978-3-031-52704-3 (eBook)
https://doi.org/10.1007/978-3-031-52704-3

This Springer imprint is published by the registered company Springer Nature Switzerland AG
The registered company address is: Gewerbestrasse 11, 6330 Cham, Switzerland

Paper in this product is recyclable.

Preface

Fouling formation is one of the most common problems in operational units of various industries, including oil, gas, petrochemical, food, pharmaceutical, power plants, etc. The fouling formation in heating equipment, in addition to destroying and damaging the devices, plays a major role in wasting energy. For this reason, many studies have been conducted regarding the experimental and theoretical investigation of the fouling formation of different heating equipment. The fouling layer formed on heat exchangers surfaces causes additional heat resistance and pressure drop due to the low conduction coefficient and the partial blocking of the flow path. As a result, the heat flux is reduced, the efficiency of the heat transfer device is reduced, and the power consumption is increased. This book is an extensive and comprehensive collection of the past experiences of the authors, as well as a collection of the latest articles in the world, which specifically deal with the formation of precipitation of salt solutions, especially calcium salts such as calcium carbonate and calcium sulfate. These salts are the most common fouling-forming salts in aqueous solutions, which have inversed solubility and their solubility decreases with increasing temperature. Therefore, in the heat exchangers where the water will experience a higher temperature, they will be precipitated. This book is the result of many years of practical and laboratory work experience of its authors, although the results of other studies have also been collected in this book in a comprehensive way.

We suggest the study of this work to managers, experts, researchers, professors, students, and those interested in the field of heat and energy transfer, and it is hoped that with the criticism and correction of these loved ones, the problems in the next editions of the book will be resolved.

Mahshahr, Iran

Mohammad Varnaseri
Seyed Mohsen Peyghambarzadeh

Contents

Abbreviations

a	Internal cross section area (m^2)
A	Heat transfer area (m^2)
A_1	Constant value in Eq. (2.2)
A_s	Cross-sectional area of shell side (m^2)
$CaCO_3$	Calcium carbonate
$CaSO_4$	Calcium sulfate
C_b	Salt concentration in bulk flow
$C_{b,\ init}$	Initial bulk concentration
C_f	Interface salt concentration
C_p	Specific heat capacity (J/g °C)
C_s	Saturated salt concentration
C^*	Interface concentration
D	Diameter
D_b	Diameter of tube-bundle
D_e	Equivalent diameter
D_h	Hydraulic diameter
D_i	Tube inner diameter
D_o	Tube outer diameter
D_s	Shell diameter
E_a	Molar activation energy for nucleation
f	Friction factor
F	Correction factor
FC	Forced convection
GDP	Gross domestic product
G_s	Shell mass velocity (kg/s m^2)
G_t	Tube mass velocity (kg/s m^2)
h	Heat transfer coefficient (W/m^2 °C)
h_{exp}	Experimental heat transfer coefficient (W/m^2 °C)
h_i	Heat transfer coefficient of inside fluid (W/m^2 °C)
h_{id}	Dirt heat transfer coefficient of inside fluid (W/m^2 °C)
h_o	Heat transfer coefficient of outside fluid (W/m^2 °C)

h_{od}	Dirt heat transfer coefficient of outside fluid (W/m^2 °C)
I	Current (A)
j_f	Friction factor
j_h	Heat transfer factor
k	Constant reaction rate in Eq. (2.3)
K	Thermal conductivity (W/m K)
K_1	Constants from Table (1.7) that used in Eq. (1.30)
k_f	Thermal conductivity of fluid (W/m K)
k_m	Mass transfer coefficient
k_{r0}	Pre-exponential constant in Eq. (2.1)
k_t	Transfer coefficient in Eq. (1.9)
kw	Thermal conductivity of the tube wall (W/m °C)
L	Tube length
l_B	Baffle spacing
$\dot{m}$	Mass flow rate (kg/s)
$\dot{m}_d$	Deposition rate
m_f	Fulling mass
$\dot{m}_r$	Deposition removal rate
$\dot{n}$	Fouling rate
n_1	Constants from Table (1.7) that used in Eq. (Chap. 1.30)
N_p	Number of tube passes
N_t	Number of tubes
Nu	Nusselt number
PB	Pool boiling
PHE	Plate heat exchanger
Pr	Prandtl number
p_t	Tube pitch
$\dot{q}$	Heat flux (W/m^2)
q	Heat transfer rate (W)
Q	Volumetric flow
R	Universal gas constant
Re	Reynolds number
R_f	Fouling resistance (m^2 K/W)
S	Distance between the thermocouples location and the surface
SFB	Subcooled flow boiling
T	Absolute temperature
t_1	Inlet temperature of cold fluid
T_1	Inlet temperature of hot fluid
t_2	Outlet temperature of cold fluid
T_2	Outlet temperature of hot fluid
T_{avg}	Bulk fluid average temperature
T_b or T_{bulk}	Bulk temperature
$T_{b,s}$	Shell side bulk fluid temperature in Eq. (1.33)
$T_{b,t}$	Tube side bulk fluid temperature in Eq. (1.33)
TEMA	Tubular Exchanger Manufacturers Association

$T_{h, in}$	Inlet hot temperature
T_{int}	Induction time
T_s or T_{surf}	Surface temperature
$T_{s, inner}$	Inner surface temperature
$T_{s, in}$	Inlet solution temperature
T_{therm}	Thermocouple temperature
T_w	Wall temperature
$t_{w, est}$	Estimated wall temperature
u	Linear or fluid velocity (m/s)
U	Overall heat transfer coefficient (W/m^2 °C)
U_c	Overall heat transfer coefficient for clean conditions
U_d or U_f	Overall heat transfer coefficient for dirty or fouled conditions
u_s	Shell side velocity (m/s)
u_t	Tube side velocity (m/s)
V	Voltage (V)
x	Fouling thickness

Greek

μ	Dynamic viscosity (Ns/m^2)
μ_w	Fluid viscosity at wall temperature (Ns/m^2)
ρ	Density (kg/m^3)
Δ_{LMTD}	Logarithmic mean temperature difference
ΔP_s	Shell side pressure drop (N/m^2 (Pa))
ΔP_t	Tube side pressure drop (N/m^2 (Pa))
ΔT_{sat}	Saturation temperature difference
ΔE	Activation energy

Subscripts

i	Inside or inner
o	Outside or outer
b	Bulk flow
f	Fouling
w	Wall
m	Metal

Chapter 1
Basic Concepts of Fouling in Heat Transfer System

Abstract One of the operational problems faced by different industries is the formation of deposits in heating equipment, which, in addition to destroying and damaging the equipment, plays a major role in energy wastage and its misuse. The deposit layer formed on the surfaces of the heat exchangers causes heat resistance and additional pressure drop due to the low conductivity coefficient and the relative obstruction of the flow path. Since the deposit layer formed on heat exchangers surfaces has a low conductivity and relatively blocks the flow path, it leads to the creation of additional heat resistance and pressure drop. As a result, the heat flux decreases, the efficiency of the heat transfer device decreases and the power consumption increases. Since more than 90% of heat transfer equipment are exposed to fouling formation, the study of fouling formation mechanisms, its prediction and control as well as the estimation of economic losses caused by this phenomenon will be very important.

Keywords Heat exchangers · Fouling formation · Fouling costs · Fouling mechanisms · Calcium salts crystallography

1.1 Introduction

For heat exchange between fluids, heat exchangers are used, which are designed and manufactured in different types such as double-pipe, shell and tube and plate heat exchangers. Heat exchangers are used in almost all cases where there is a need for heat exchange (in industry and household). The main use of this equipment is in large industries such as oil, gas and petrochemical, power generation, electricity, power plant, wastewater treatment industries, etc. In this way, the entire market of heat exchangers is valued at billions of euros, and with the increase in the world's population and energy demand, this market is expected to expand even more. Almost the presence of particles and salts in suspended or dissolved form in most fluids is a proven fact, during the heat exchange of these fluids in heat transfer equipment, there will be a possibility of their accumulation due to temperature changes. Accumulation of these materials, which are known as fouling, leads to problems such as increasing

M. Varnaseri and S. M. Peyghambarzadeh, *Scale Formation in Heat Exchangers*,
https://doi.org/10.1007/978-3-031-52704-3_1

pressure drop, reducing heat transfer capacity, changing flow geometry and surface topography in most heat exchangers (more than 90% of heat exchangers face these problems [1]).

1.2 Fouling Formation Importance in Heat Exchangers

Of the 1.36×10^{18} m^3 of water on earth, ocean water makes up more than 97%. In 1970, about 42% of the water used for cooling in the US power industry was ocean water. This water contains various mineral ions. The most important of them is the Ca^{+2} ion, which forms crystalline deposits in weakly alkaline environments [2]. Calcium as the most common metal ion is found in different concentrations in natural water [3]. Salt (crystalline) fouling formation on heated surfaces due to their inverse solubility with temperature is one of the challenges of various industries [4–10]. This deposit, which is formed by the accumulation of unwanted materials during a time process, creates an additional layer with low thermal conductivity on heated surfaces. Deposit layer creates resistance, reduces performance and economic efficiency of the heat transfer system, accumulates in pipelines, orifices and other flow channels, blocks them and increases the pressure drop [11–20]. Energy loss, excess energy consumption, increased cleaning costs, reduced equipment capacity, increased corrosion, metal erosion, process fluid contamination, reduced equipment safety level, reduced equipment life, increased metal surface temperature, and improper flow distribution are some of the problems that fouling creates [6–8, 21]. The formation of fouling reduces the performance of heat exchangers by 80% and sometimes causes their complete destruction. Therefore, it is very important to identify the fouling processes and operational parameters effective in increasing the amount of deposition [22]. One of the factors that complicate the fouling formation process is the simultaneous existence of two or more fouling formation mechanisms [23]. The general point of view in heat exchangers design to deal with fouling is an average of 35% oversize for the heat exchanger, which makes the heat exchangers heavier, larger and more expensive [13]. A common practice in designing heat exchangers is to use a fixed value for fouling resistance (R_f). This constant value can predict the final amount of deposition in heat exchangers, but it cannot predict the time of its occurrence. Therefore, it may be necessary to take the heat exchanger out of service at an unfavorable time for cleaning and removal of deposits [24]. In the early 1950s, the first set of fouling resistances were published in the Tubular Exchanger Manufacturers Association (TEMA) standards. Although TEMA values still form the basis of the most heat exchangers design worldwide, there are several problems associated with the use of TEMA fouling resistances, such as:

- Their origin and operational conditions are not known.
- Most of the values are related to water or hydrocarbon flow.
- They apply only to shell and tube heat exchangers.

- They do not provide any information on operational parameters effects such as flow rate, fluid composition, heat flux and temperature on fouling rate.
- It is not clear after which operating time the fouling resistance was obtained.
- These data do not provide time-dependent management of fouling resistance.

There are two types of salts: (1) Salts that precipitate on cold surfaces (known as salts of normal solubility) such as $CaCl_2$ and NaCl; (2) Salts that precipitate on heated surfaces (known as salts of inverse solubility) such as $CaCO_3$ and $CaSO_4$ [25]. Therefore, for optimal design, the effect of different operating conditions on $CaCO_3$ and $CaSO_4$ crystallization fouling should be thoroughly investigated [26].

The environmental problems of heat exchanger fouling have recently received a lot of attention due to the problems related to the emission of CO_2, NO_x and SO_x and carcinogenic waste, the limitation of the use of fouling chemical inhibitors, the limitations of using cooling water and increase in temperature, and the limitation of chemical waste disposal [27]. The steps taken during the heat exchanger design are the most effective steps in controlling fouling. These steps include choosing the heat exchanger optimal type, choosing the optimal operating conditions (higher velocities, lower temperatures, etc.) and choosing the optimal heat exchanger design. Gilmour [28] reported that shell and tube exchangers fouling mainly results from poor shell-side design, which will ultimately lead to reduced equipment performance. Compared to shell and tube exchangers, plate exchangers have received more attention due to less breakdown [29]. If the factors affecting deposition are identified, it can be prevented or its amount can be reduced [12]. In addition, it is possible to reduce the costs caused by fouling by identifying the mechanisms governing it and controlling these mechanisms [1]. Process fluid characteristics and their constituents (for example, solid particle presence) can cause deposits formation under the influence of operating conditions [4]. The influencing factors on crystallization fouling are divided into two groups of process conditions (including salt system, supersaturation, pH, flow rate, flow regime and additives) and interface conditions (including temperature, surface energy, roughness and topography surface, amount of nucleation spots, aging of fouling layer and surface) [9]. Due to the dependence of fouling on factors such as flow rate, fluid characteristics, heated surface temperature, geometry and material, heat transfer rate and, etc., it is not possible to generalize the data of fouling tests to other systems [19]. The parameters that engineer consider when designing heat exchangers to overcome the problem of fouling are (a) Velocity: high flow velocity disrupts the formation of deposit and reduces its amount. (b) Bulk fluid temperature: the chemical reaction of the fouling is affected by this factor because it determines the sufficiency of the activation energy for a chemical reaction. (c) Heat exchanger surface temperature: this temperature controls the solid materials formation rate on the surface. (d) pH: determines the acidity or alkalinity suitable for the formation of some minerals. (e) Surface material: some surfaces are prone to biological fouling growth and some also promote mineral fouling [10]. In another classification, factors affecting the formation of fouling are divided into three general categories (a) operational parameters, (b) fluid properties and (c) heat exchanger parameters [17]. Due to the wide range of factors affecting fouling, in this book only operational parameters including fluid flow rate, surface temperature, bulk temperature, etc. have been discussed. Förster et al. [30] have stated that advanced antifouling strategies are based on approaches that increase the induction period length. These strategies include two approaches to improve the heated surface properties (include energy and geometry) in order to increase the induction period length and adjust the flow hydrodynamic conditions to minimize the fouling occurrence. In this book, the operational factors affecting the crystallization fouling of $CaSO_4$, $CaCO_3$ and composition of these salts have been investigated. Our goal is to create general ideas among crystallization fouling research and to understand the effects of different factors on this type of fouling. In addition, the weak and strong points of previous researches have been revealed in the review and the possibility of creating new research areas has been identified. In this book, in order to better understand the concepts and compare the results of different researches, various data from the articles available in the last few decades (more than 50 years) have been collected and the results have been compared.

1.2.1 Heat Exchangers

As can be seen in Fig. 1.1 (adapted from the study of [31]), heat exchangers are classified from different points of view such as pass and flow arrangement, flow regime, heat transfer mechanism, etc.

1.2.2 Fouling Costs for Industries

The cost that fouling has imposed on the economy of industrialized countries is close to 0.25% of Gross Domestic Product (GDP) [32, 33]. Müller-Steinhagen [34], in a study conducted in 1999, presented the annual costs that fouling has imposed on the economy of several large industrialized countries in Table 1.1. The costs related to fouling formation can be divided into four general categories [35]:

(a) Capital expenditures: These costs include the allocation of more surface area, the cost of a stronger foundation due to larger equipment, increased costs of transportation, installation and larger space, costs of anti-fouling equipment including on-line cleaning systems installation, pre-treatment units and cleaning equipment on-site.
(b) Fuel expenditures: If fouling causes more fuel consumption in boilers or furnaces, or more energy such as electricity or steam, fossil fuels and other resources are needed to overcome fouling effects, these costs are included in the fuel group.

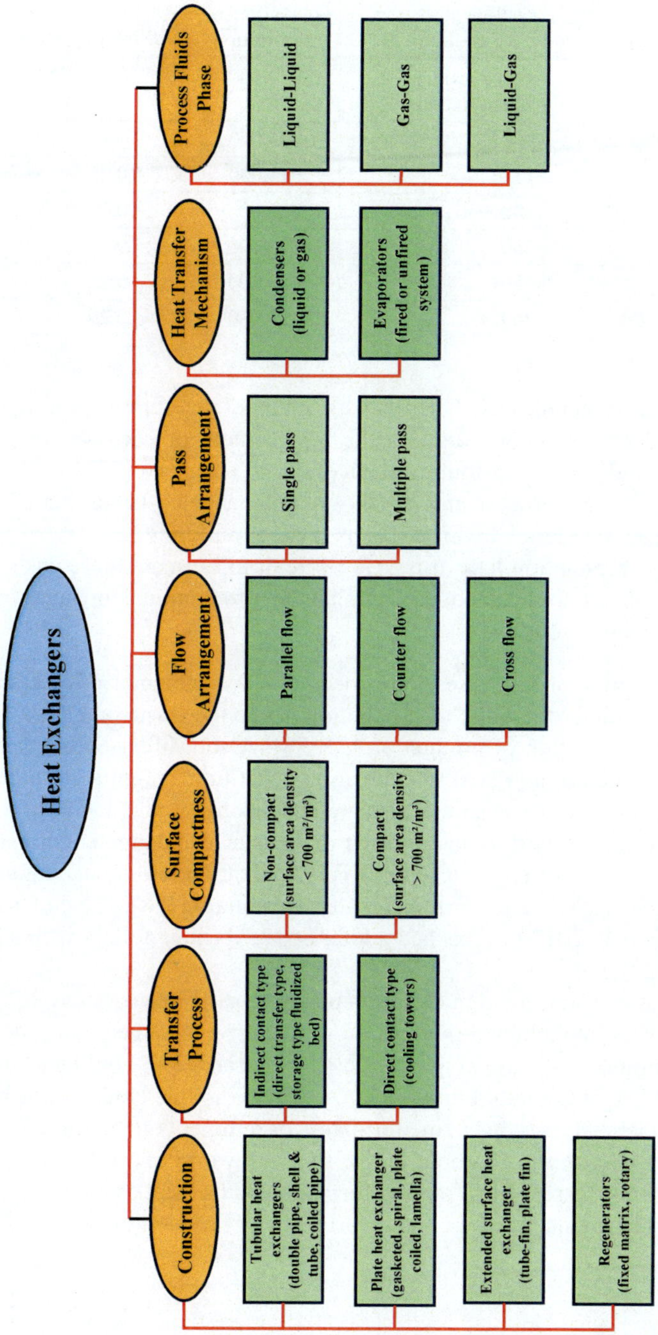

Fig. 1.1 Heat exchangers classification (adapted from [31])

Table 1.1 Estimation of annual fouling costs in some countries [34]

Country	Fouling cost US $ million	GDP (1984) US $ million	Fouling costs % of GDP
USA (1982)	3860–7000 8000–10,000	3,634,000	0.12–0.22 0.28–0.35
Japan	3062	1,225,000	0.25
Germany	1533	613,000	0.25
UK (1978)	700–930	285,000	0.20–0.33
Australia	260	173,000	0.15
New Zealand	35	23,000	0.15
Total industrial world	26,850	13,429,000	0.20

(c) Maintenance expenditures: These costs include the removal of deposited fouling, the cost of chemicals and the replacement of corroded or plugged equipment and other anti-fouling device costs. Thackery [27] and Pritchard [36] stated that in a process unit, 15% of maintenance costs are related to heat exchangers and boilers, of which 50% is related to fouling.

(d) Costs due to production loss: It is very difficult to estimate these costs. These costs are related to planned and unplanned production stoppages of heat exchangers due to fouling.

The fouling formation increases the processes environmental effects and also reduces the economic profitability of production due to the waste of energy and raw materials and the reduction of production [21, 25]. Also, Müller-Steinhagen et al. [37] showed that the costs related to deposition constitute a significant part of the current costs of the industry. The use of plate heat exchangers is increasing due to their higher heat transfer performance, easier maintenance and greater compactness, but their performance is often reduced due to crystallization fouling due to insoluble salts deposition on heated surfaces in supersaturated solution [38]. The global market of heat exchangers in 2012 has reached a total of 12.7 billion dollars with a growth of 3–5% per year.

In addition, due to unwanted materials deposition on heated surfaces that occurs in most exchangers, manufacturers focus on producing exchangers with more heat recovery conditions and the use of more efficient materials [39]. Fouling is the emission factor of 2.5% of the total human-related carbon dioxide emissions across the globe [21]. Efficiency losses due to the formation of sediment on the surfaces of heat exchangers are estimated to be about 2% of the total world energy production per year [40]. Efficiency losses due to fouling formation on the heat exchangers surfaces are estimated to be about 2% of the total world energy production per year [40].

1.3 Heat Exchangers Problems in the Presence of Fouling

As mentioned in the previous sections, the formation of deposits leads to the appearance of problems such as pressure drop, reduction of thermal efficiency, sub-fouling corrosion, etc., and the final result of all these issues is the failure of the heat exchanger. Corrosion is actually the mechanical deterioration of the heat exchanger structural materials due to the fluid in the flow and the environment, which is in contact with it. In addition, the deposition phenomenon is mainly in transition conditions and as a result of all three phenomena of momentum, heat and mass transfer. Besides corrosion, other mechanical phenomena are also important, such as erosion, which is the same corrosion at the place of contact and metal surfaces, which basically occurs when the part is under load or exposed to vibrations and sliding. In general, fouling and corrosion reduce the heat exchanger efficiency and should be considered in the heat exchanger design [41]. Formation of deposits on heat transfer surfaces is one of the basic problems in heat transfer and boiling with water. Due to operational problems and unrealistic design methods, heat exchangers are usually designed with a factor of 70–80% higher than the actual surface (overdesign), of which 30–50% is considered for fouling. Considering the additional heat transfer surface may increase the heat exchangers operating time, but it cannot have an effect on the fouling.

The heat exchangers fouling can be reduced by its correct design and the correct selection of the heat exchanger type and mechanical and chemical mitigation methods. It is necessary to determine the type of fouling mechanism and the dependence of fouling rate and operational factors in all its mitigation methods. It is still not possible to predict the fouling resistance in terms of time in a specific composition of water, but the effects of water velocity, concentration of the precipitating element and the contact surface temperature on the fouling rate can be estimated and used to optimization in design of heat exchanger.

Due to the low thermal conductivity of these depositions (Table 1.2), their thermal resistance increases, and with the reduction of heat exchanger performance, depending on the amount of deposition, periodic cleaning becomes necessary. The increase in the tube inner surface temperature at a constant thermal load for a reboiler only with the formation of a deposit with a thickness of 8 mm has been around 112 °C, and this temperature increases at contact surface of tube and deposit has increased the rate of fouling formation and corrosion [42].

From another point of view, the fouling effects can be classified into three main sections including technical, ecological and economic, which is shown in Fig. 1.2. According to Fig. 1.2, the most important negative effect of fouling is the increase in the overall industrial operations cost and the depreciation of equipment efficiency, which leads to the loss of profit from production.

Table 1.2 Thermal conductivity for different cooling deposits [43]

Deposit type	k (W/m K)
Sodium aluminum silicate	0.2–0.4
Hematite (boiler fouling)	0.6
Biofilm	0.7
Calcium sulfate (boiler)	0.8–2.2
Calcite (boiler)	0.9
Serpentine (boiler)	1.0
Gypsum (boiler)	1.3
Calcium sulfate	2.3
Magnesium phosphate	2.3
Calcium phosphate	0.6
Calcium carbonate	0.6
Magnetic iron oxide	0.9
Composition of milk	0.5–0.7

1.3.1 Environmental Problems of Fouling

Formation of deposits in heat transfer systems leads to more pressure drop, improper flow distribution, significant loss of heat transfer and sub-fouling corrosion. The consequences of these unwanted phenomena will be increased costs, environmental, safety and health problems. In general, the negative effects of fouling can be summarized as follows:

(1) Shut-down of operational units, which leads to a decrease in production and an increase in costs.
(2) Increasing maintenance costs in operational units.
(3) Decrease in efficiency and operational capacity and as a result decrease in production.
(4) Cost increases due to chemical antifouling used, or mechanical antifouling equipment, or the destruction of the equipment and the need to replace it.
(5) Increased initial design cost due to the need for additional heat transfer surface.
(6) Increasing production of greenhouse gases and other environmental risks due to production effluents caused by the use of antifouling. (Environmental problems such as water/land pollution, increased production of carbon dioxide, SO_x, NO_x, production of hazardous waste, use of chemical inhibitors, etc.)

Note: a preliminary estimate by Müller-Steinhagen et al. [21] have shown that approximately 88 million tons of carbon dioxide per year (equivalent to 2.5% of the total anthropogenic carbon dioxide emissions) are released into the environment due to the formation of fouling in crude oil refineries. In addition, the heavier the crude oil, the more complicated its processing and the amount of carbon dioxide discharge to the environment will increase.

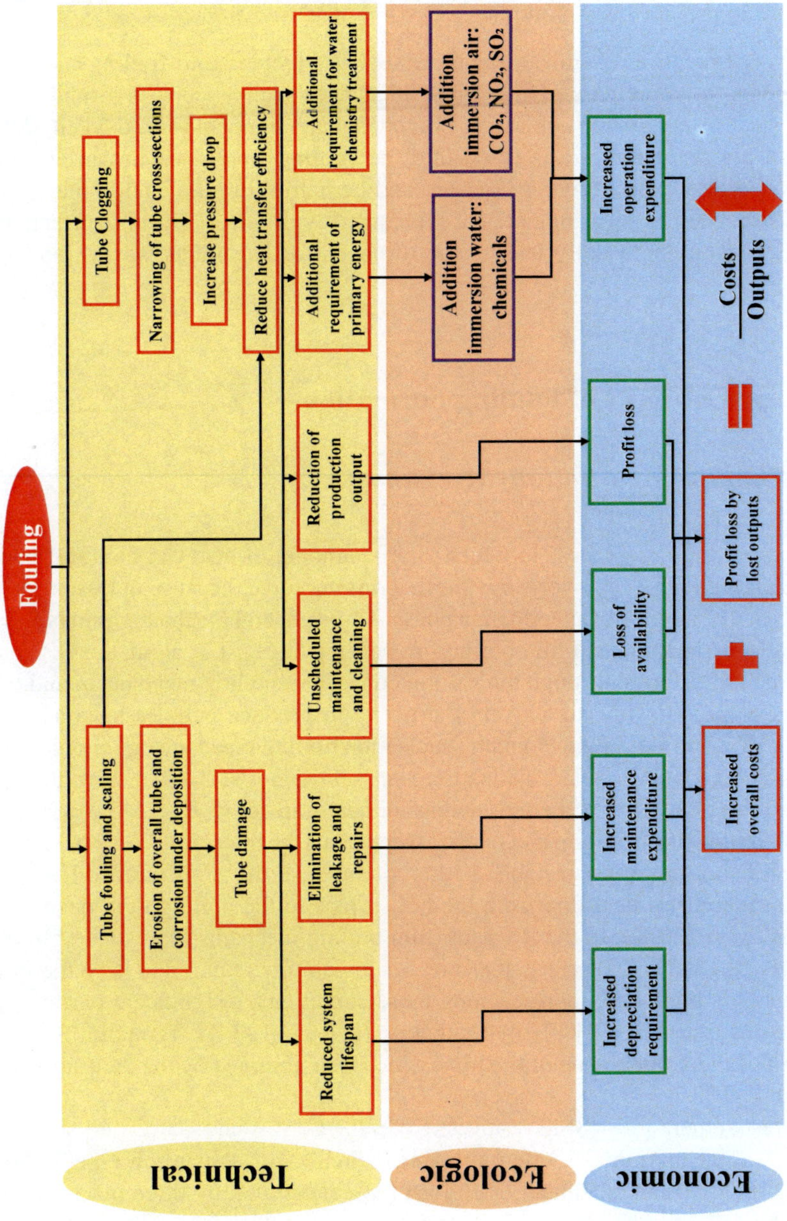

Fig. 1.2 Overview of fouling effects in industries (adapted from [39])

(7) Safety risks during the cleaning operation of heat exchangers.
(8) Increase in the cost of consumption resources such as water, electricity, fuel and etc.
(9) Failure to provide design output.

Various chemicals as chemical inhibitors such as biocides, anti-fouling and anti-scalant agents, in combination with other substances such as corrosion inhibitors, additives such as coagulants, chlorine, polyphosphate, hypochlorite, oxygen scavengers and pH regulators are injected into the cooling water; which can eventually penetrate into water resources or land and be harmful to the environment and humans. Also, chemical washing of heat exchangers with acids, solvents, surfactants and alkaline solutions usually lead to the discharge of hazardous wastes into the environment.

1.4 Basic Concepts of Fouling Formation

1.4.1 Fundamental Boiling Mechanisms

Nucleation boiling is a process in which a large amount of heat can be transferred from the surface with a relatively low thermal driving force, because in this process the liquids at the nucleation sites change phase. Three general mechanisms in boiling are nucleation, transition and film boiling. According to Fig. 1.3, along the OP line, heat has been transferred through the natural convection boiling mechanism and the conditions of superheating to activate the nuclei to produce bubbles have not yet occurred ($\Delta T_{sat} < 5$ K). In the PQ path (nucleation boiling mechanism), bubbles are formed at certain points called nucleation sites on the hot surface, the production of these bubbles causes mixing and increasing heat transfer (5 K $< \Delta T_{sat} <$ 30 K). The QR path (transition boiling mechanism) represents the transition from nucleation boiling to film boiling and is unstable. In this condition, as the temperature difference increases, the transfer heat flux from the hot surface to the boiling fluid decreases (30 K $< \Delta T_{sat} <$ 120 K). In the RS path (film boiling mechanism), a stable film of vapor covers the hot solid surface, the surface temperature required to keep the film boiling stable is high, and upon reaching these conditions, a significant part of the surface heat is transferred by thermal radiation ($\Delta T_{sat} >$ 120 K). The point S is the upper limit of the boiling zone of the film, which is determined by the melting point of the metal [44].

Many researchers [45–48] have investigated the $CaSO_4$ and $CaCO_3$ crystallization, as well as the mixture of these salts, using recirculating tubular flow type fouling measuring devices. Jamialahmadi et al. [49] have stated that by three mechanisms heat transfer occurs in boiling liquid, heat transferred by force convective to liquid, conduction by liquid microlayer evaporation between bubble and heated surface and heat transfer from the superheated liquid boundary layer to the vapor bubbles. Of these three mechanisms, only micro-layer evaporation significantly increases the

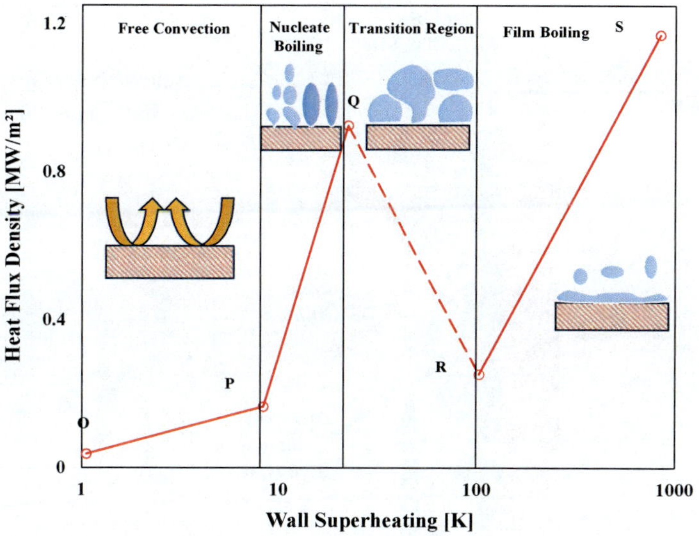

Fig. 1.3 General boiling diagram for water (adapted from [44])

concentration of the precipitating substance under the bubble. They have also stated that one of the most severe forms of deposition occurs during pool boiling (Fig. 1.4) heat transfer (such as kettle reboilers and steam generators) [18]. Subcooled flow boiling occurs in some industrial applications such as boilers, boiling water reactors and a new generation of electronic and computer systems. This type of boiling occurs near the heated wall while the bulk fluid is in subcooled conditions (below the saturation temperature). Bubbles burst rapidly if they are removed from the developing saturated layer [50]. The fouling process is divided into two periods: "induction period", which is characterized by nucleation and crystal growth stable on the heated surface; and the "growth period": which is characterized by compact fouling layer formation that leads to incremental and measurable fouling resistance with time [11, 15]. The induction period in which stable nucleation and crystal growth take place on the heat transfer surface, the fouling growth period in which the increase in fouling resistance can be measured as time increases due to the formation of a compact deposit layer [11, 15]. In addition, the deposition formation stages on surface during pool boiling process are:

(a) Bubble nuclei formation above the nucleation sites: Since the bubble density is proportional to the number of nucleation sites, the material and surface roughness are very important.
(b) Agitation and disturbance caused by bubbles: This phenomenon causes a significant increase in the heat transfer coefficient. The characteristics of the boiling liquid, especially the surface tension, affect the size, shape and bubbles rising speed.

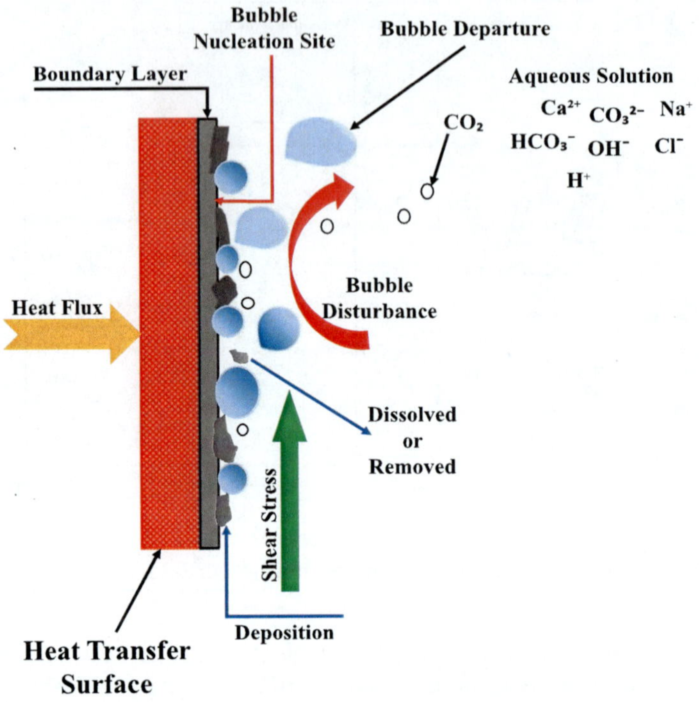

Fig. 1.4 Schematic of the formation of vapor bubbles formation, motion, fouling deposition and fouling removal on the vertical surface in pool boiling (adapted from [51])

(c) Micro layer formation under the bubbles: Faster evaporation of the solvent increases the concentration of the dissolved salt, the temperature gradient in the microlayer causes the supersaturation of the solution, in this condition the crystal nuclei are formed in the form of concentric rings around the nucleation sites [15, 18].

1.4.2 Fouling Mechanisms

Based on physical and chemical mechanisms, deposits are usually divided into six groups. These groups include biological, particulate, chemical reaction, crystallization, corrosion, and freezing (Solidification), and this classification is also shown in Fig. 1.5 [4, 5, 12, 17, 33]. Crystallization accounts for up to 25% of all fouling problems among the mentioned mechanisms [6, 32]. This type of deposit, which occurs due to the dissolved species precipitation from process solution on heated surfaces, on various industries worldwide has the most detrimental effect [25]. Crystallization fouling is a complex phenomenon that depends on chemical kinetics, thermodynamics and material properties in addition to hydrodynamic and thermal conditions

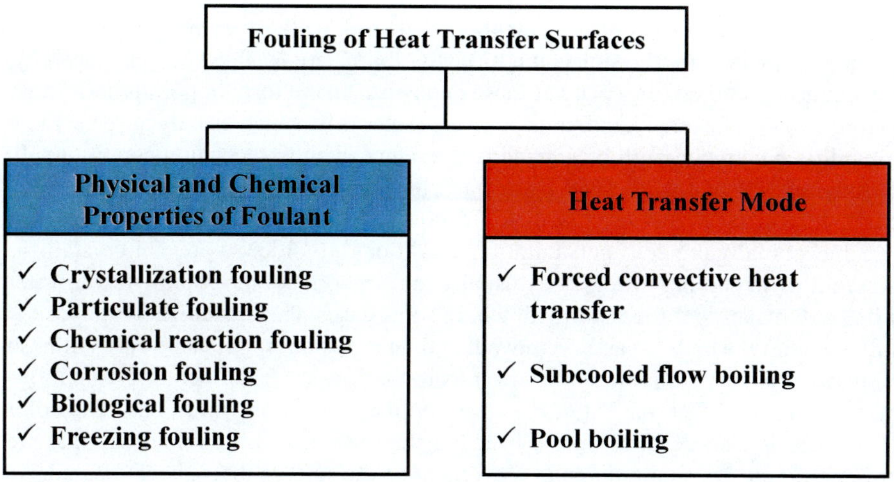

Fig. 1.5 Fouling classification based on the physical/chemical properties of the precipitants and the type of heat transfer (adapted from [55])

of the system [52]. The behavior of the same salts in different systems and operating conditions is very different, this makes it necessary to carefully investigate the behavior of fouling formation in different systems and operating conditions [53]. Crystallization fouling includes three basic stages:

(a) Supersaturation: The ratio of the salt concentration in bulk flow (C_b) to saturated salt concentration (C_s), and is obtained by solution evaporating beyond the salt solubility limit with inverse solubility properties.
(b) Nucleation: A process in which the smallest stable mass is formed from a crystallized phase in a crystal system, and then these particles act as centers of crystallization. Nucleation includes two types, homogeneous and heterogeneous. In heat exchangers, nucleation is generally achieved by a foreign body such as particles with the fluid and is heterogeneous. Since the presence of particles in industrial fluids is inevitable, this can be a serious problem [29].
(c) Crystal growth: According to Mullin's view [54], two mechanisms of transporting ions from bulk flow to interface of liquid/crystal and surface ions integration in crystal lattice cause crystals growth created during the nucleation stage.

1.4.2.1 Crystallization Fouling

Crystal formation of water-soluble salts on heat transfer surfaces is known as crystallization or scaling. The creation of supersaturation conditions as the driving force of crystallization is necessary for the formation of this type of fouling. The existence of a small amount of supersaturation degree determines the formation of fouling, as well as the temperature distribution between the heated wall temperature and bulk fluid the place of formation. As mentioned before, salts of normal solubility are precipitated when, firstly, supersaturation conditions are established, and secondly, the solution is cooled (the shape of these deposits is in the form of porous and slime-like layers). On the other hand, salts of inverse solubility, which usually have very low solubility, precipitate with increasing temperature in supersaturation conditions. In cooling towers, heat is removed from water through partial evaporation in the spray pond, this partial evaporation leads to the concentration of salts dissolved in water, which eventually create supersaturation conditions. In order to prevent crystallization fouling in cooling towers, they usually replace some of the cooling tower water with fresh make-up through blow-down. In some cases, the crystallization deposit is soft and slimy, which is easily removed, and on the contrary, in some cases, it is so hard that it is very difficult to remove it from the surface. The operating conditions of the contact surface have a direct effect on the deposition conditions. The color of deposition also depends on the heat transfer system and the impurities that are present. Figure 1.6 shows the fluid movement on the surface that is being heated. As Fig. 1.6 shows, if concentration driving force includes bulk fluid salt concentration (C_b) and interface salt concentration (C_f), the fouling process is controlling by mass transfer, and if concentration driving force includes interface concentration (C_f) and saturation concentration (C_s), the fouling process is controlling by chemical reaction [30].

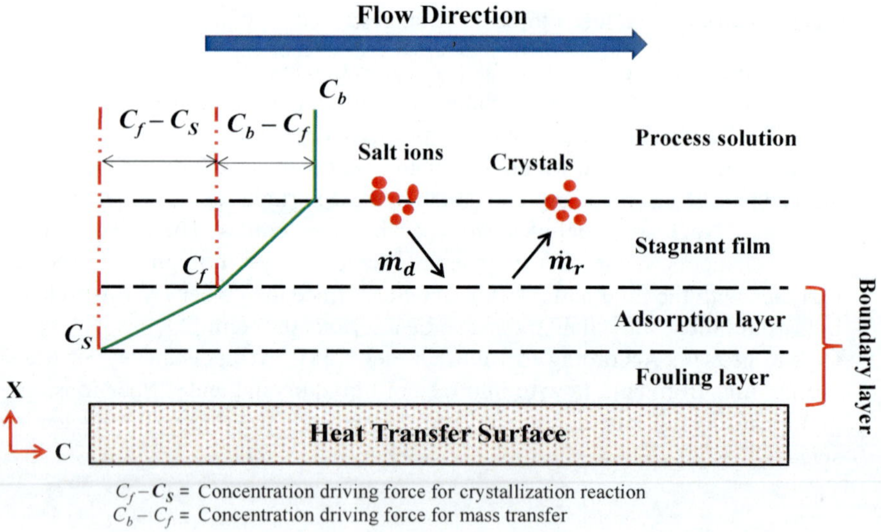

Fig. 1.6 Different of mechanisms in crystallization fouling (adapted from [30])

1.4.2.2 Particulate Fouling

The placement or transfer of suspended particles from the fluid on the heated surface leads to the formation of particulate fouling. These suspended particles are different depending on the heat transfer system. Some particles may be naturally present in the fluid, or due to temperature increasing and chemical reaction, the particles are created in the form of polymer molecules and suspended in the fluid. Mineral particles in river water, soot particles due to improper combustion, salt deposition in desalination systems, suspended particles in cooling tower water, dust particles in air coolers are examples of particulate fouling. Some influencing factors on this type of fouling are: flow rate, heat flux, particle concentration, heated surface temperature.

1.4.2.3 Biological Fouling

It is defined as the growth of living matter on heat exchanger surfaces, usually caused by microorganisms. The creation of this type of fouling, which is due to the growth of large microorganisms or organisms on the heated surface, causes problems in water flows. Usually, this type of fouling is divided into two categories: macrobial and microbial. If the deposition is caused by the accumulation of marine vegetation or shells (macro-organisms), it is called microbial deposit; On the other hand, if bacteria, fungi, algae, etc. accumulate on the heated surface (micro-organisms), it is called microbial deposit. This type of fouling usually occurs in single-pass systems or open cooling water systems (micro-organisms such as microbes, bacteria and algae need light to grow) and appears in the form of a biofilm. Their removal is difficult due to their unevenness, flexibility, etc., and generally, due to the activity of these living organisms, the pH of the deposit area decreases and leads to sub-deposition corrosion in the heat transfer surface. Algae and fungi are usually not on the surface of heat exchangers, but their presence in the cooling fluid can lead to other deposits or flow blockages downstream.

1.4.2.4 Chemical Reaction Fouling

In some cases, a chemical reaction on the heated surface may lead to the production and accumulation of deposits. Usually, the increase in temperature is the key factor in increasing the rate of chemical reactions, and as a result, the probability of the reaction at the heat transfer surface is much higher. Therefore, chemical reaction fouling can occur in a wide temperature range and in all types of process fluids. A very common example of this fouling is the milk pasteurization process. Removal of chemical reaction fouling is directly dependent on the type of chemical reaction involved.

1.4.2.5 Corrosion Fouling

In this type of deposit, a layer that has a lower thermal conductivity replaces the heat transfer surface that has disappeared spontaneously. The reason why many metal alloys are resistant to corrosion is the presence of a very thin oxide layer on their surface, which prevents corrosion by limiting the flow of ions and electrons, which are the main cause of corrosion. This protective layer usually has a very thin thickness so that it does not interfere with heat transfer. If the surface corrosion is high, such as in the presence of acids (aggressive chemical agents), this protective layer (oxide layer) itself causes fouling. The materials from which the heat exchanger is made are the determining factors in the formation of corrosion deposits, on the other hand, sometimes the cost of using corrosion-resistant alloys is so high that it is necessary to limit and control the corrosion of metals in another way.

1.4.2.6 Freezing (Solidification) Fouling

If the heat transfer surface reaches a temperature lower than the freezing point of the fluid that is in contact with it, solidification (freezing) fouling occurs. Usually, the formation of paraffin wax precipitation by cooling hydrocarbons is known as freezing fouling. Another example of this type of fouling is when the temperature of the heat transfer surface on which the water flows fall below zero degrees Celsius, as a result of which a layer of ice is inevitable on that surface. This type of deposit can be removed by choosing the temperature of the cooling liquid (the temperature of the heat transfer surface should be controlled above the freezing point of the cooling liquid).

1.4.2.7 Mixed Fouling Mechanisms

Real deposits in industrial heat transfer equipment usually do not include only one fouling type, but a combination of several fouling mechanisms. For example, in the case of cooling water, in addition to micro-organisms (biological fouling agent), suspended particles (particulate fouling agent), dissolved solids (crystallization fouling agent) or the presence of acidic chemicals (corrosion or chemical fouling agent) are likely to exist. Therefore, we are faced with fouling composed of several mechanisms that make the fouling process and the method of dealing with it very complicated. On the other hand, when a corrective action is taken to eliminate this type of fouling, all fouling should be considered with different mechanisms.

1.5 Formulation to Determine Fouling Resistance and Net Fouling Rate

Figure 1.7 shows the layers of deposits around a heated surface, which are of different resistances. For this heated surface, the overall energy balance is expressed as follows:

$$q = Ah_i(T_{bi} - T_{fi}) = A\frac{k_{fi}}{X_{fi}}(T_{fi} - T_{wi}) = AU_c(T_{wi} - T_{wo}) \quad (1.1)$$

$$q = Ah_o(T_{fo} - T_{bo}) = A\frac{k_{fo}}{X_{fo}}(T_{wo} - T_{fo}) = AU_f(T_{bi} - T_{bo}) \quad (1.2)$$

where

q heat transfer rate;
A heat transfer area;
h heat transfer coefficient;
k thermal conductivity;
X thickness;
U_c overall heat transfer coefficient for clean conditions;
U_f overall heat transfer coefficient for fouled conditions.

In clean conditions, the overall heat transfer coefficient given in relation (1.1) is defined as follows:

$$\frac{1}{U_c} = \frac{1}{h_i} + \frac{X_m}{k_m} + \frac{1}{h_o} \quad (1.3)$$

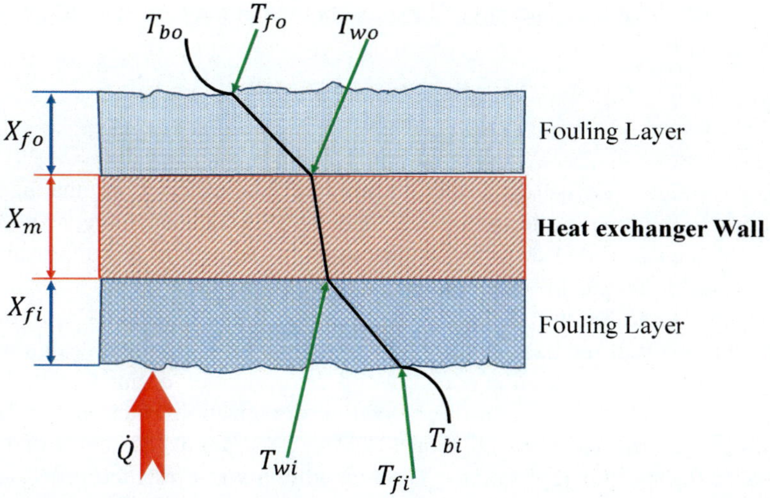

Fig. 1.7 Temperature profile in the deposited heat exchanger wall (adapted from [42])

in deposited conditions, the overall heat transfer coefficient is expressed in the form of the following equation:

$$\frac{1}{U_f} = \frac{1}{h_i} + \frac{X_{fi}}{k_{fi}} + \frac{X_m}{k_m} + \frac{X_{fo}}{k_{fo}} + \frac{1}{h_o} \tag{1.4}$$

if it is assumed that the deposition occurs only on one side of the heated surface (almost in most cases this assumption is correct), Eq. (1.4) becomes as follows:

$$\frac{1}{U_f} = \frac{1}{h_i} + \frac{X_m}{k_m} + \frac{X_f}{k_f} + \frac{1}{h_o} \tag{1.5}$$

and fouling resistance is defined by Eq. (1.6):

$$R_f = \frac{1}{U_f} - \frac{1}{U_c} = \frac{X_f}{k_f} \tag{1.6}$$

by using Eq. (1.7), it is possible to determine the mass of deposition per unit area of deposit layer:

$$m_f = \rho_f k_f R_f \tag{1.7}$$

Assuming that the thermal conductivity and density of the deposited layer are constant, it can be said that the fouling resistance (R_f) is proportional to the fulling mass (m_f). Fluid flow on the surface usually result in a shear force on the surface, when the deposit is formed on the surface, this shear force will lead to the removal process. This removal process is intensified due to the fouling layer becoming more fragile in the upper fouling layer thicknesses or suspended particles presence in fouling crystal lattice. Equation (1.8) expresses the net increase rate of the fouling layer [42]:

$$\frac{dm_f}{dt} = \dot{m}_d - \dot{m}_r \tag{1.8}$$

The "d" refers to deposition and "r" to removal, where dm_f/dt, $\dot{m}_d$ and $\dot{m}_r$ are the net fouling formation rate, deposition rate, and deposition removal rate, respectively. As can be seen in Fig. 1.8, different processes of deposition and removal occur simultaneously in typical fouling system.

Using the same fouling factors for the gas–gas heat exchanger shows that the investment cost will increase slightly. In the traditional method of designing heat exchangers, the issue of fouling potential and determination of appropriate fouling resistance are related. To aid in this selection, organizations such as the Tubular Exchanger Manufacturers Association (TEMA) have published tables of fouling resistances for specific applications. The first edition was presented in 1941, since then the data has been revised. The main problem in reaching the design is the issue of choice. In the best fouling resistance tables, a range of resistances is given, but

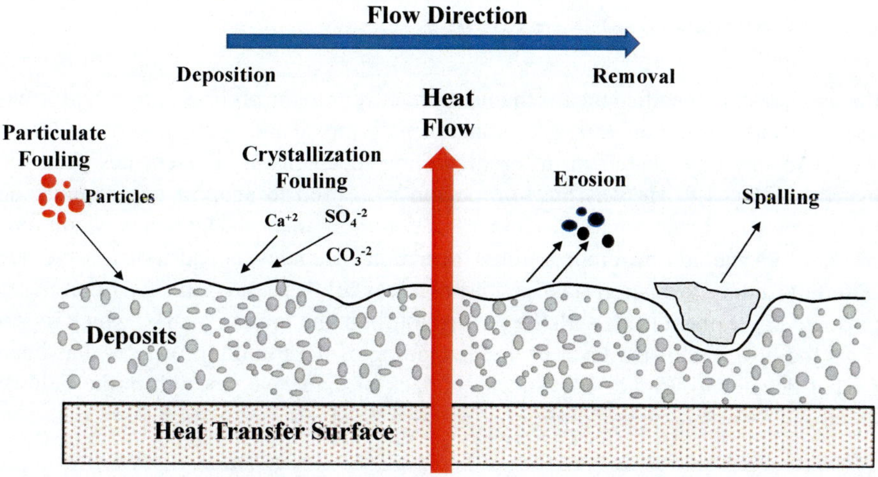

Fig. 1.8 Various processes of deposition and removal in a typical fouling system (adapted from [12, 56])

generally there is no information on which value to use. For example, in general, there is no information about fluid rate, temperature, nature and concentration of precipitating substances. Later, it was seen that other factors can have an obvious effect on fouling resistance progress. Probably most amount information in tables is related to water. The data in Table 1.3 has been published by TEMA based on careful revision and engineering vision for operation and design of complex shell and tube heat exchangers.

Table 1.3 Fouling resistances in the water system [42]

Water type	Fouling resistance (10^4 m^2 K/W)
Sea water (outlet max 43 °C)	1.75–3.5
Brackish water (outlet max 43 °C)	3.5–5.3
Treated cooling tower water	1.75–3.5
Artificial spray pond (outlet max 43 °C)	1.75–3.5
Closed loop treated water	1.75
River water	3.5–5.3
Engine jacket water	1.75
Distilled water or closed cycle condensate	0.9–1.75
Treated boiler feed-water	0.9
Boiler blow down water	3.5–5.3

1.6 Sequential Events in Fouling

Epstein [57] has specified the mechanisms that occur in the process of fouling formation, including initiation, transport, attachment, removal and aging. In order to check the frequency of research in different fouling mechanisms, Epstein designed the matrix of Fig. 1.9. He was the first person who tried to analyze and explain the fouling process by summarizing it in a 5 × 5 fouling matrix. The 5 matrix columns in Fig. 1.9a include different fouling formation mechanisms and its rows include different fouling formation stages (sub-processes) of these mechanisms. Figure 1.9b presented by Bohnet [58]; with the passage of time, the fouling matrix, which represents the progress of the level of understanding of the fouling processes, has been further developed. Figure 1.9c shows the lack of extensive research in the field of aging and removal stages, despite conducting a lot of research in the field of transport stage.

1.6.1 Initiation

With the start of the test, it takes a long time until a significant deposit is observed on the clean heat transfer surface, this time is called the delay period or initiation. An increase in heated surface temperature and supersaturation degree (these two factors are related to each other) lead to an increase in crystal nucleation phenomenon, which in turn lead to a decrease in the delay period in crystallization fouling. In addition, the increase in surface temperature leads to the intensification of the reactions induced in the chemical reaction fouling, which ultimately reduces the delay period [59]. The role of fluid flow velocity is different and depends on other factors [60]. Heat transfer surface characteristics have a significant effect on delay period, such that rough ridges (roughness) lead to increased nucleation sites, surface chemical activities and adsorption, while grooves lead to an increase in the deposition area. The transfer

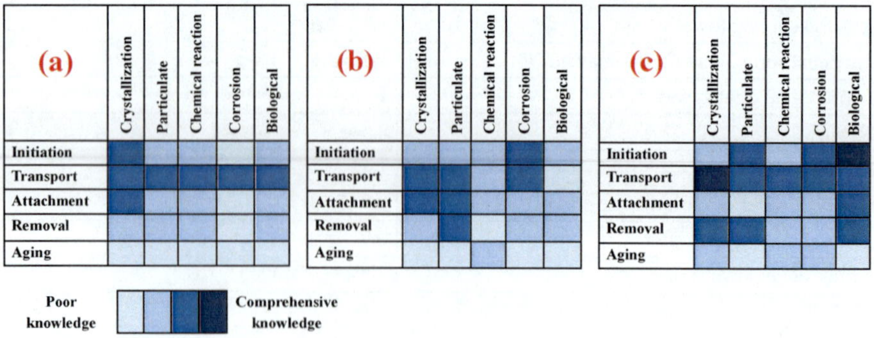

Fig. 1.9 General views of the 5 × 5 matrix of fouling mechanisms (adapted from [57])

of the eddy towards the heated surface is also one of the consequences of reducing the thickness of the viscous sub-layer due to the increase in surface roughness. In addition, Baier [61] has stated that the surface adsorption of proteoglycans is the first step in biofouling (these materials are always present in natural water) that microorganisms adhere to them in order to form surface conditioning films.

1.6.2 Transport

In the fouling process, the transport stage is the most understandable stage. At this stage, different deposit species or critical reactants or even oxygen must be transferred from bulk fluid (where concentration is equivalent to C_b) to the heated surface (where concentration in the fluid adjacent to the surface is equivalent to C_s). Equation (1.9) expresses this phenomenon known as local fouling flux ($\dot{m}_d$):

$$\dot{m}_d = k_t(C_b - C_s) \tag{1.9}$$

where k_t is the transfer coefficient. In the case of ions, molecules or sub-micrometer particles, the transport is diffusive in nature, and k_t is equivalent to the known mass transfer coefficient k_m.

1.6.3 Attachment

After transferring the deposits from bulk fluid to heated surface along the boundary layer, the deposits connect to the heated surface as well as to themselves. In general, surface factors including surface energy, the nature and composition of previous layers, as well as operational factors such as fluid flow velocity (or shear force on the surface) and surface temperature strongly affect the attachment. Then salt ions that approach the surface gradually grow over time. Packter [62], in a study on the sequential events of nucleation and fouling growth, stated that under the forces of electromagnetic, deposits to form a nucleus started to stick to heated surface and eventually deposition accumulated on surface.

1.6.4 Removal

Usually, the result of deposition and removal of precipitation substance until the growth of the fouling layer on the heated surface becomes uniform is steady as fouling. In contrast to the relatively good understanding of deposition mechanisms, removal mechanisms are not well known [63]. The main factor in removing fouling from the heated surface is the shear force near the interface, this shear force is influenced by various factors such as the velocity gradient at the interface of the heated surface/fouling, fluid viscosity and surface roughness.

1.6.5 Aging

As fouling begins, aging begins. In the aging stage, under higher surface temperature influence for a long period of time, the fouling layer shows the physical or chemical transformation of the crystals, which changes the properties and structure of fouling. The mechanical strength changes in the accumulation of fouling due to the change of mechanical properties is caused by the change of crystals. In industrial operations, the aging stage plays an important role by managing planning for cleaning strategies.

1.7 Fouling Curve

Fouling process is indicated by fouling resistance (R_f), which can be measured experimentally [64]. Fouling resistance (R_f) is obtained from Eq. (1.10) [65].

$$R_f = \frac{1}{h(t)} - \frac{1}{h(t=0)} \tag{1.10}$$

where h represents heat transfer coefficient and t represents time, Eq. (1.10) shows that the fouling resistance is obtained from the difference between the heat transfer coefficient at the beginning of each test and the heat transfer coefficient after a specific operating period.

If the graph of fouling resistance (R_f) is plotted against time for crystallization fouling, four time periods representing three main regions will be identified. The first region is where the fouling resistance becomes zero and then negative, at first, there is no deposit (the fouling resistance is zero), then nucleation begins, the presence of nuclei roughens the surface and the initial growth of the fouling layer diffuse under the viscous sub-layer and creates turbulence [66], as a result, the convective heat transfer coefficient increases (fouling resistance becomes negative), this is the nucleation phase (induction or delay time). With the formation of more deposit, the thermal resistance increases and overcomes the advantage of turbulence, and as a result, the fouling resistance becomes positive. The period of time from the beginning of the test to when the fouling resistance becomes positive after being negative for some time is called the roughness delay time [29]. After that there is a region where the fouling resistance increases steadily, this is the growth region.

The growth of crystals is intensified and a deposit layer is formed on the surface. Finally, there is a region when the fouling resistance does not change due to the equality of the growth and removal rates of the deposit layer, this phase is asymptotic. These three regions when considered together reveal an "S" shaped curve [5, 15, 25, 29]. Induction time is divided into nucleation delay time and roughness delay time. These two periods are affected by unordered nucleation and the growth of the first crystals on the heat transfer surface. Many factors such as process parameters (including layer forming component concentration, temperature, flow regime and velocity, pH value, solubility behavior), operational parameters (including cleaning, fluctuation in process conditions, flow regime, heat flux) and equipment parameters (apparatus design, geometry, material selection, surface properties) affect the formation and period length [12, 46, 67]. After the roughness delay time, the fouling curve can be classified into three different categories according to Fig. 1.10 for constant heat flux conditions [55]:

(a) Linear: A linear fouling curve is obtained where either the removal rate is negligible (strong fouling) or is constant, but less than the deposition rate, i.e., the net fouling rate is constant.
(b) Falling: The falling curve is obtained for deposits with lower mechanical resistance. Accumulation of deposition increases the removal rate with time, as a result, the net fouling rate decreases.
(c) Asymptotic: Asymptotic curve occurs for weak deposits where the removal rate increases with time and eventually equals the deposition rate. Then the net fouling rate becomes zero.

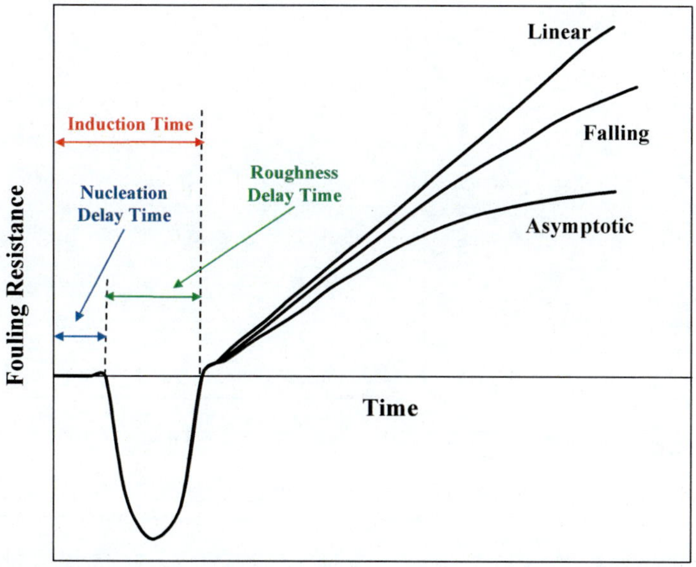

Fig. 1.10 Types of fouling curves (adapted from [25])

1.8 Change in Fouling Thickness Over Time

Three basic stages may be seen in relation to the formation of fluid deposits moving on surfaces:

- The transfer of precipitating materials or precursors from the flowing fluid to the vicinity of the boundary layer of the solid surface,
- Attachment of deposits to the surface or themselves,
- Transfer of matter out of the surface.

The sum of these rules shows the growth of fouling on the surface. The mathematical expression of fouling growth rate depends on the difference between $\dot{m}_d$ and $\dot{m}_r$, which are the deposition rate and the removal rate, respectively. The amount of attachment will be affected by $\dot{m}_r$. Figure 1.11 shows an idealized asymmetric curve of fouling growth rate on the surface. In region *A*, the induction period is shown. In some situations of fouling formation, the induction period can be long, maybe up to several weeks. Region *B* shows the stable growth of fouling on the surface. Under these conditions, there will be a competition between deposition and removal. The deposition rate gradually decreases, while the removal rate increases gradually. Finally, the removal rate and the deposition rate may equalize, until in region *C*, when the actual thickness of the deposit remains constant, the flat or asymptotic steady state is reached. Many system variables have extensive effects on different stages and this issue will be discussed in later chapters.

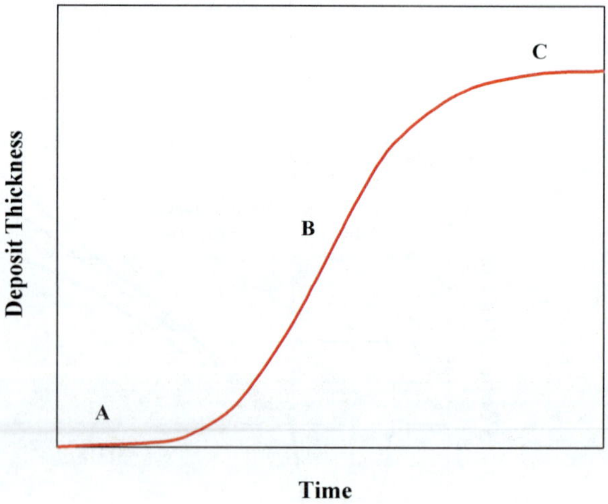

Fig. 1.11 Change in fouling thickness over time (adapted from [42])

1.9 Introduction of Salt Deposits

1.9.1 Crystallography of Calcium Salts

As mentioned earlier, calcium salts account for a large proportion of mineral deposits in heat transfer systems [11]. The crystallography of some of these deposits is described below.

1.9.1.1 Calcium Carbonate ($CaCO_3$)

One of the hard and resistant mineral deposits that can be found abundantly in nature is calcium carbonate ($CaCO_3$). This salt has three anhydrous polymorphs including vaterite, calcite and aragonite [68]. The crystal structures are presented in Fig. 1.12 [69, 70]. Among calcium carbonate polymorphs, it has been shown that vaterite has the lowest and calcite has the highest thermodynamic stability [71]. Calcite usually forms at room temperature and has a hexagonal crystal shape, while aragonite belongs to the orthorhombic system and irreversibly transforms into calcite under the conditions of heating in dry air to about 40 °C. The transformation rate increases with increasing temperature. When aragonite is in contact with water or solutions containing calcium carbonate, transformation is possible even at room temperature. Vaterite, on the other hand, is less stable and transforms into calcite and aragonite under geologically dependent conditions, although it can also be observed during high-temperature crystallization from calcium carbonate [72].

1.9.1.2 Calcium Sulfate ($CaSO_4$)

Calcium sulfate is one of the most common deposit constituents. As a solid, calcium sulfate crystallizes from an aqueous solution in three forms: gypsum ($CaSO_4 \cdot 2H_2O$), calcium sulfate hemihydrate ($CaSO_4 \cdot 0.5H_2O$), and calcium sulfate anhydride ($CaSO_4$) [73]. The crystal structures are shown in Fig. 1.13 [74]. Gypsum is both the primary material before dehydration and the main product after rehydration and usually precipitates in the range of 40–98 °C, while the hemihydrate and anhydride are species that probably precipitate at temperatures above 98 °C [75].

1.9.1.3 Co-precipitation of Calcium Carbonate and $CaSO_4$

At the end of the twentieth century, very little attention was paid to the phenomenon of co-precipitation [76, 77]. Now, with the help of developed experimental and analytical tools, more studies have been developed in this field [78]. For the co-precipitation

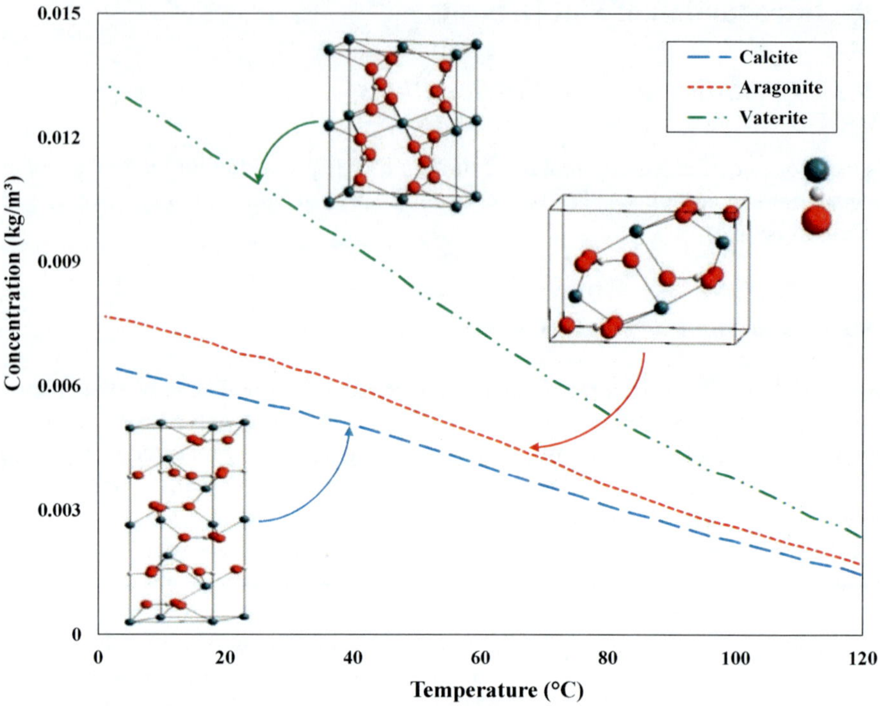

Fig. 1.12 Solubility of different forms of calcium carbonate as a function of temperature and concentration. Crystal structure: C, O and Ca atoms are represented by gray (smallest), red (largest) and dark green (medium size) spheres, respectively (adapted from [65, 69, 70])

of calcium carbonate and $CaSO_4$, some qualitative studies have determined that $CaSO_4$ crystals tend to grow on the surface of calcium carbonate [79]. The presence of $CaSO_4$ weakens the calcium carbonate precipitate, making it less hard and easier to move in solution, while the co-precipitation is much more sticky than pure $CaSO_4$.

1.9.2 *Calcium Salts Solubility*

1.9.2.1 Calcium Carbonate

Calcium carbonate crystals exist in three forms: vaterite, calcite and aragonite. Since all three forms of these salts have inverse solubility, their solubility in water decreases with increasing temperature. The solubility of different calcium carbonates as a function of temperature is shown in Fig. 1.12 [65]. Najibi et al. [65] have shown that

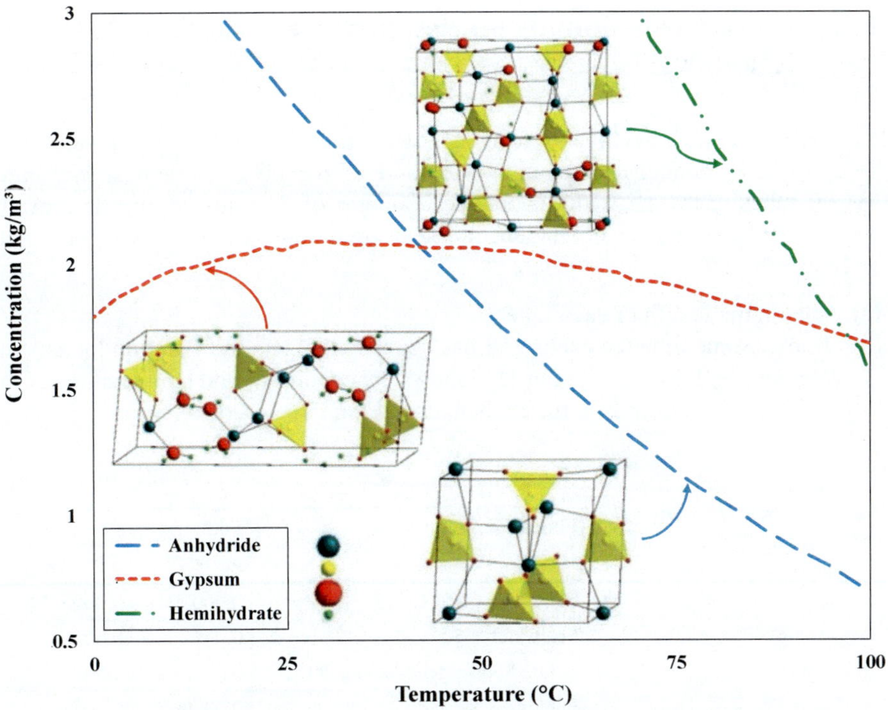

Fig. 1.13 Solubility of different forms of calcium sulfate as a function of temperature and concentration. Crystal structure: H, S, O and Ca atoms are represented by green (smallest), yellow (second smallest), red (largest) and dark green (second largest) spheres, respectively. The surface (yellow) represents the group (adapted from [74, 80])

more than 99% of calcium carbonate deposits are in the form of aragonite in subcooled flow boiling. Of the calcium carbonate polymorphs, calcite is thermodynamically more stable, while aragonite may change to calcite. Over time, all crystals eventually transform to calcite at any temperature.

1.9.2.2 Calcium Sulfate

The solubility of calcium sulfate in water as a function of operating temperature is shown in Fig. 1.13 [80]. At a temperature higher than 40 °C, the solubility of calcium sulfate decreases with increasing temperature. Calcium sulfate precipitates appear from a water solution in three forms: gypsum ($CaSO_4 \cdot 2H_2O$), calcium sulfate hemihydrate ($CaSO_4 \cdot 0.5H_2O$), and calcium sulfate anhydride ($CaSO_4$). The reaction rate of the anhydride, which is the most thermodynamically stable form, is so slow that only the other two forms are usually observed. The solubility of all three forms decreases with increasing temperature.

1.10 Considering Fouling on Design Stage (Practical Examples)

Example 1 A Shell and tube exchanger E-101 is used to heat 7.5 kg/s of water from 85 to 99 °C by condensing steam at a pressure of 345 kN/m^2. The heat exchanger has 60 tubes each one with an outside diameter of 2.5 cm. During the design, the overall heat transfer coefficient in clean conditions was used with the value of 2800 W/m^2 °C.

(a) Obtain the length of each tube.
(b) If after some time the exchanger has deposited, obtain the water outlet temperature using Table 1.4. Figure 1.14 shows the conditions and temperatures of the inlet and outlet fluids to the exchanger of Example 1-part a.

Table 1.4 Selected fouling factors [81]

Fluid type	Fouling factor	
	m^2 °C/W	h ft^2 °F/Btu
Sea water < 50 °C	0.00009	0.0005
Sea water > 50 °C	0.002	0.001
Treated boiler feedwater > 50 °C	0.0002	0.001
Fuel oil	0.0009	0.005
Quenching oil	0.0007	0.004
Alcohol vapors	0.00009	0.0005
Steam, non-oil-bearing	0.00009	0.0005
Industrial air	0.0004	0.002
Refrigerating liquid	0.0002	0.001

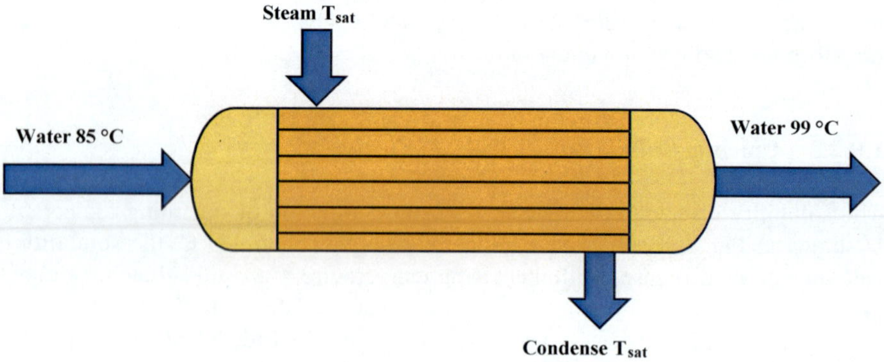

Fig. 1.14 Input and output fluid temperatures and conditions related to the heat exchanger Example 1-part a.

- According to the given pressure ($P = 345\ kN/m^2 = 3.45$ atm), the saturation temperature (T_{sat}) is obtained from thermodynamic tables; which is equal to 138 °C.
- Now the Logarithmic Mean Temperature Difference (Δ_{LMTD}) is calculated:

$$\Delta_{LMTD} = \frac{(T_1 - t_2) - (T_2 - t_1)}{\ln \frac{(T_1 - t_2)}{(T_2 - t_1)}} = \frac{(138 - 99) - (138 - 85)}{\ln \frac{(138-99)}{(138-85)}} = 45.64\,^{\circ}\mathrm{C} \quad (1.11)$$

- On the other hand, if there is only a phase change in the heat exchanger, F = 1.

$$P = \frac{t_2 - t_1}{T_1 - t_1} = \frac{99 - 85}{138 - 85} = 0.264 \quad (1.12)$$

$$R = \frac{T_1 - T_2}{t_2 - t_1} = \frac{138 - 138}{95 - 85} = 0.0 \quad (1.13)$$

from Fig. 1.15:

$$\rightarrow F = 1.$$

$$q = \dot{m}C_p\Delta T = 7.5 \times 4180 \times (99 - 85) = 438{,}900\ \mathrm{W} \quad (1.14)$$

$$q = UAF\Delta T_{LM} \rightarrow A = \frac{438{,}900}{2800 \times 1 \times 45.64} = 3.43\ \mathrm{m}^2 \quad (1.15)$$

$$A = \pi D_o L N_t \rightarrow L = \frac{3.43}{\pi \times 0.025 \times 60} = 0.73\ \mathrm{m} \quad (1.16)$$

- From Table 1.4: R_f (water): 0.0002 m^2 °C/W; R_f (steam): 0.00009 m^2 °C/W;
- The sum of the resistances (Fig. 1.16):

$$\sum R_f = R_{f\,(\text{tube side})} + R_{f\,(\text{shell side})} = 0.0002 + 0.00009 = 0.00029\ \mathrm{m}^2\,^{\circ}\mathrm{C/W} \quad (1.17)$$

$$\frac{1}{U_d} = \frac{1}{U_c} + \sum R_f = \frac{1}{2800} + 0.00029 \rightarrow U_d = 1545\ \mathrm{W/m^2\,^{\circ}C} \quad (1.18)$$

From Eqs. (1.14) and (1.15):

$$q = \dot{m}C_p\Delta t = 7.5 \times 4180 \times (t_2 - 85);$$

$$q = UAF\Delta T_{LM} = 1545 \times 3.43 \times 1 \times \frac{(138 - 85) - (138 - t_2)}{\ln \frac{(138-85)}{(138-t_2)}}$$

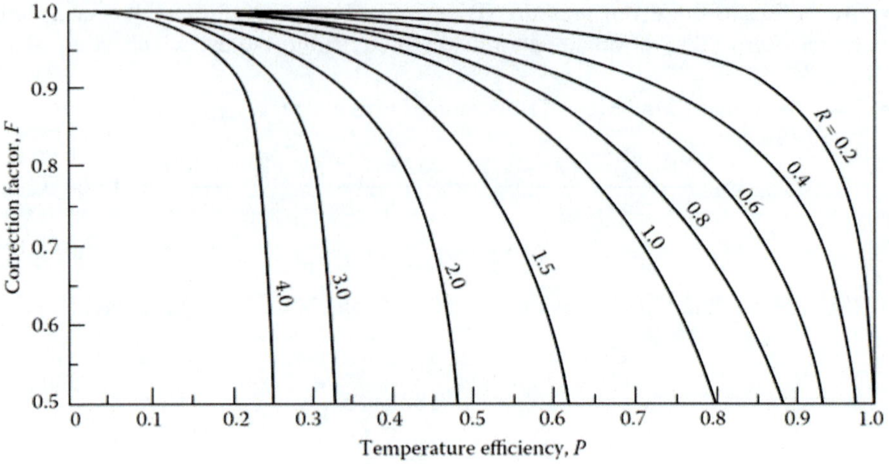

Fig. 1.15 LMTD correction factor F for a crossflow heat exchanger with both fluids unmixed [82]

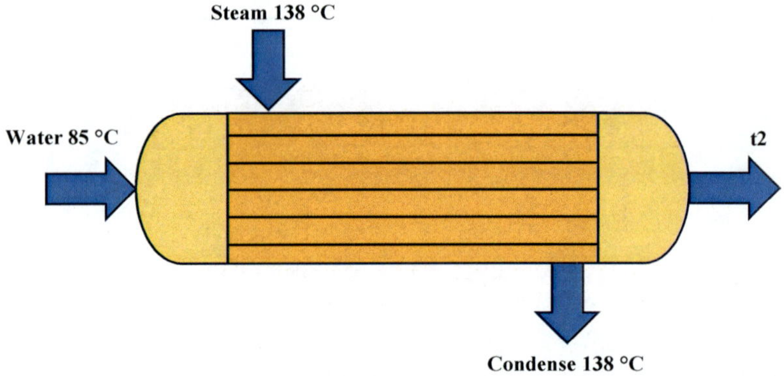

Fig. 1.16 Input and output fluid temperatures and conditions related to the heat exchanger Example 1-part b.

- We must use an iterative procedure to determine the outlet temperature: t_2 = 93.22 °C.

So, it made sense that in the beginning, the design was based on U_{dirty}; and therefore, the length of the tubes increased. For this purpose, it is necessary to use the bypass path for the initial start-up times so that the temperature does not exceed 99 °C (Fig. 1.17).

Example 2 In the laboratory work of our group [83], an experimental system was used to find the heat transfer coefficient and fouling factor according to Figs. 1.18 and 1.19. The detailed description of this experimental system is clearly presented

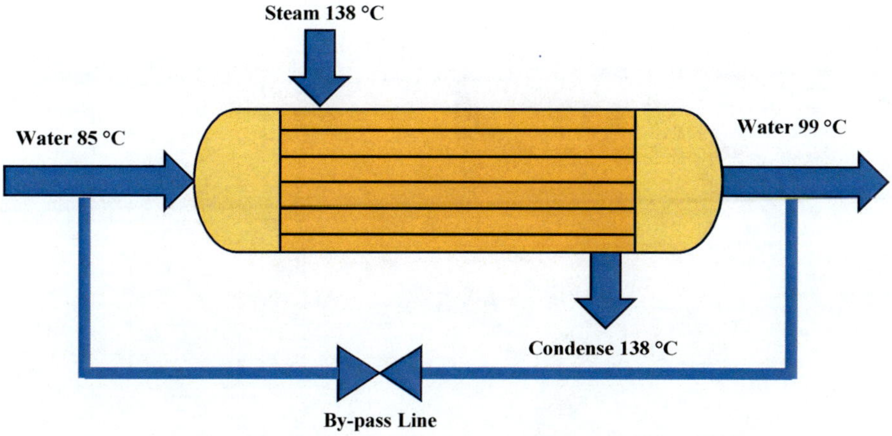

Fig. 1.17 Bypass path suggestion for initial start of exchanger in Example 1-part b.

in our articles [1, 19, 50, 56, 83–92]. In the following, the method of obtaining the mentioned coefficients in one of the experiments are described.

At first, the method of calculation and determination of operating parameters in the experimental system is explained.

The surface temperature (T_{surf}) is calculated as the arithmetic mean of the temperatures read from 4 thermocouples (T_{therm}).

$$T_{surf} = \frac{T_{therm1} + T_{therm2} + T_{therm3} + T_{therm4}}{4} \tag{1.19}$$

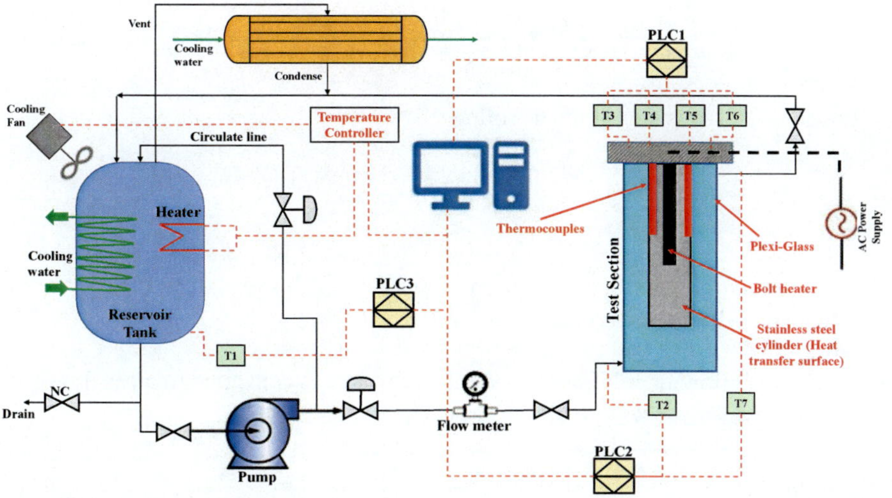

Fig. 1.18 Test apparatus [83]

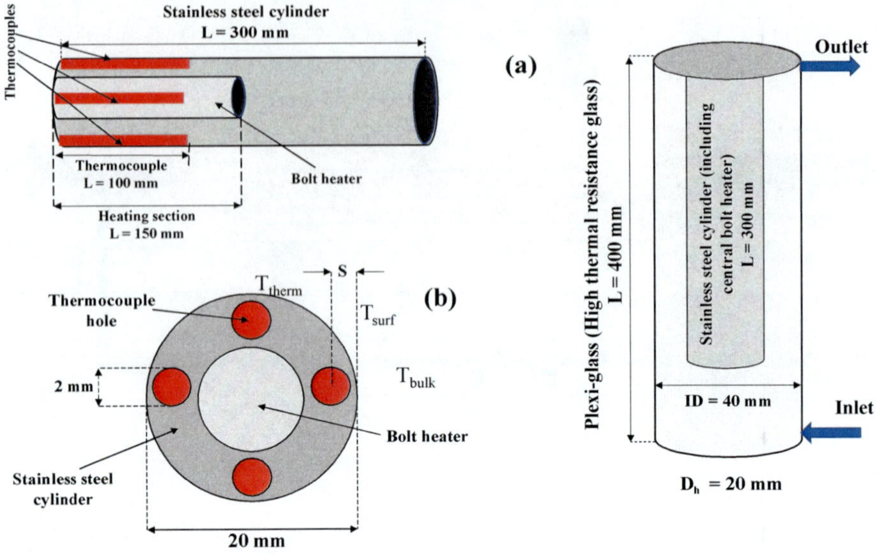

Fig. 1.19 **a** Graphical shape of test section, **b** test heater cross section and the thermocouples' locations [83]

Due to the distance between the thermocouples and the surface, the temperature indicated by them differs slightly from the actual value; This difference is due to the heat conduction resistance of the heater material which is located between these two points. These temperatures can be related to each other according to the energy balance in steady-state condition.

$$q = U(T_{therm} - T_{bulk}) = h(T_{surf} - T_{bulk}) = \frac{K}{S}(T_{therm} - T_{surf}) \tag{1.20}$$

Equation (1.20) is simplified as follows:

$$\frac{1}{U} = \frac{1}{h} + \frac{S}{K} \tag{1.21}$$

- Finally, according to the assumptions and relationships explained in detail in the articles [84–92], the average value of 5×10^{-4} W/m^2 K is considered for S/K term.
- Using the following relationship, volume flow rate is converted to mass flow rate:

$$\dot{m} = \rho \times Q \tag{1.22}$$

- The desired heat flux is obtained by changing the voltage and from the following equation:

$$q = V \times I/A \tag{1.23}$$

$$A = \pi \times D_h \times L = (3.14 \times 0.02 \times 0.15) = 9.42 \times 10^{-3}\ m^2 \tag{1.24}$$

The heat transfer coefficient can be calculated using a simple energy balance as follows:

$$h \cdot A \cdot (T_{surf} - T_{bulk}) = \dot{m} \cdot C_p \cdot (T_{out} - T_{in}) \tag{1.25}$$

$$h_{exp} = \frac{\dot{m} \cdot C_p \cdot (T_{out} - T_{in})}{A \cdot (T_{surf} - T_{bulk})} = \frac{\dot{q}}{(T_{surf} - T_{bulk})} \tag{1.26}$$

The fouling factor is obtained from the difference between the heat transfer coefficient before the start of the test ($t = 0$) and the heat transfer coefficient at any desired time according to Eq. (1.10).

In one of the tests, the fluid flow rate was 1.5 l/min, the solution contained 550 ppm of calcium carbonate salt, the amount of applied voltage and ampere were 100 V and 1.5 A, respectively. After starting the set-up and reaching the steady-state condition, the thermocouples temperatures as well as the inlet and outlet bulk temperatures have been read as follows:

$$T_{therm\,1} : 86.8\,°C;\ T_{therm\,2} : 87\,°C;\ T_{therm\,3} : 89\,°C;$$
$$T_{therm\,4} : 90.9\,°C;\ T_{in} : 60\,°C;\ T_{out} : 61.5\,°C$$

$$\rightarrow T_{surf} = 88.425\,°C \text{ and } T_{bulk} = 60.75\,°C$$

- At the bulk temperature (T_{bulk}), the physical properties of the fluid are calculated:

$$C_p = 4180.80\,(J/kg\,K);\ \rho = 982.77\,\left(kg/m^3\right)$$

- From Eq. (1.14)

$$\dot{q} = \frac{\dot{m}C_p\Delta T}{A} = \frac{(\rho Q)C_p\Delta T}{A} = 16{,}356.6\,\frac{W}{m^2}$$

- Through the specified voltage and ampere, the applied heat flux is calculated and compared with the heat flux calculated from Eq. (1.23); If the average error value is reasonable, the calculations continue.

$$\text{Eq. (1.23)}: q = V \times I = 100 \times 1.54 = 150\ W;$$

$$\dot{q} = \frac{q}{A} = \frac{154}{0.0094} = 16{,}382.9\ W/m^2$$

$$AAD = \left|\frac{\dot{q}_{\exp} - \dot{q}_{cal}}{\dot{q}_{\exp}}\right| = \left|\frac{16{,}353.6 - 16{,}382.9}{16{,}353.6}\right| = 0.18\% \tag{1.27}$$

$$\text{Eq. (1.26)}: h_{\exp} = \frac{\dot{q}}{T_{surf} - T_{bulk}} = \frac{16{,}356.6}{(88.425 - 60.75)} = 591.02\ \text{W/m}^2\ \text{K}$$

Now, after the desired time has passed, the calculations are repeated in the same way and a new value for the experimental heat transfer coefficient is calculated. Next, the fouling resistance at the desired moment is obtained from Eq. (1.10). For example, when the value of experimental heat transfer coefficient is 582.38 W/m^2 K at about 10 min after the start of the experiment, the fouling resistance is

$$\text{Eq. (1.10)}: R_f = \frac{1}{582.38} - \frac{1}{591.02} = 0.00002\ \text{m}^2\ \text{K/W}$$

As can be seen, the value of fouling resistance changes depending on the operating conditions and also with the passage of time, and putting a fixed value in the design of heat exchangers according to the TEMA standard can be problematic. Therefore, the effect of operating parameters on the fouling resistance should be investigated with more extensive studies and this issue should be considered during the design of heat exchangers.

Example 3 In this example, Kern's method is used, which was obtained based on an experimental work in the field of commercial heat exchangers and predicts the heat transfer coefficient in an acceptable way. A real example of the heat exchanger design is presented and the role of the fouling factor is stated at the beginning of the design. It should be noted that in this method, mass and linear velocities are considered according to the change of the cross-sectional area along the shell diameter based on the maximum level the cross flow.

Cooling of 100 tons/h of methanol from 95 to 40 °C is done by cooling water. The water inlet temperature is 25 °C and the outlet temperature is 40 °C [93].

- Due to the corrosiveness of water, it is placed on tube side, and the hot fluid methanol is placed inside the shell. The arrangement of the tubes is considered to be triangular considering the cleanliness of the fluid on the shell side. At the average fluid temperature, their physical properties are as Table 1.5.
- Heat load

Table 1.5 Physical properties at average temperature of fluids [81]

Properties	Methanol	Water
C_p (kJ/kg °C)	2.84	4.2
μ (mNs/m^2)	0.34	0.8
K_f (W/m °C)	0.19	0.59
ρ (kg/m^3)	750	995

$$\text{Eq. (1.14)}: \ q_h = \dot{m}_h C_{p_h}(T_1 - T_2) = \frac{100{,}000}{3600} \times 2.84(95 - 40) = 4340\ \text{kW}$$

$$\dot{m}_C = \frac{q}{C_{p_c}(t_2 - t_1)} = \frac{4340}{4.2(40 - 25)} = 68.9\,\frac{\text{kg}}{\text{s}}$$

$$\text{Eq. (1.11)}: \ \Delta T_{LMTD} = \frac{(T_1 - t_2) - (T_2 - t_1)}{\ln \frac{(T_1 - t_2)}{(T_2 - t_1)}}$$

$$= \frac{(95 - 40) - (40 - 25)}{\ln \frac{(95-40)}{(40-25)}} = 31\,^{\circ}\text{C}$$

The Ft value (deviation from the counter-current flow) is calculated according to Fig. 1.20 by determining the values of R and P.

$$R = \frac{(T_1 - T_2)}{(t_2 - t_1)} = \frac{95 - 40}{40 - 25} = 3.67 \tag{1.28}$$

$$P = \frac{(t_1 - t_2)}{(T_1 - t_1)} = \frac{40 - 25}{90 - 25} = 0.21 \tag{1.29}$$

$$\rightarrow \text{Ft} = 0.85$$

$$\rightarrow \Delta T_{LM} = F_t \Delta T_{LMTD} = 0.85 \times 31 = 26\,^{\circ}\text{C}$$

where

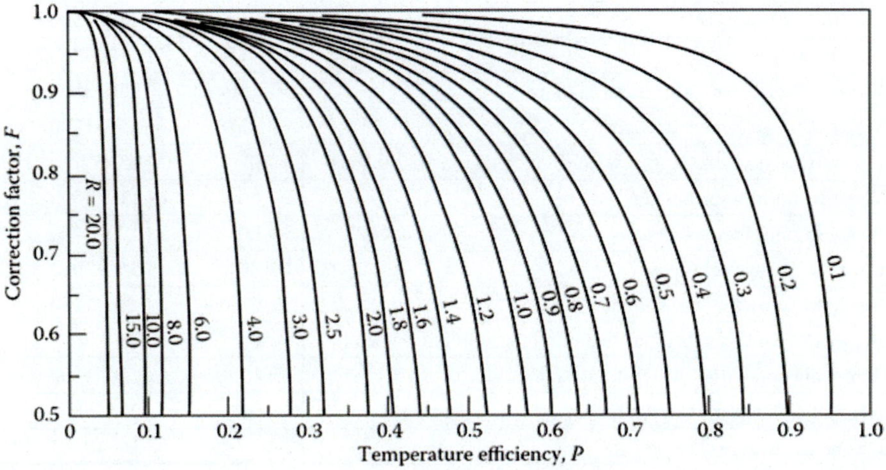

Fig. 1.20 Correction-factor plot for heat exchanger with one shell pass and two, four, or any multiple of tube passes [82]

ΔT_{LM} Corrected temperature difference.

- From Table 1.6, an initial guess is determined for the overall heat transfer coefficient under clean condition, which here is 600 W/m^2 °C.
- The overall external surface of the tubes (from Eq. 1.14):

$$\text{Eq. (1.14)}: \; A = \frac{q}{U \Delta T_{LM}} = \frac{4340 \times 10^3}{600 \times 26} = 278 \text{ m}^2$$

- According to the TEMA standard, section 5, the outer diameter of the tubes is 20 mm and their length is 16 ft (4.87 m).
- External surface area of a tube:

$$A = \pi DL = \pi \times \left(20 \times 10^{-3}\right) \times (4.87) = 0.303 \text{ m}^2$$

Table 1.6 Approximate values of overall heat-transfer coefficients [81]

Physical situation	U	
	Btu/h ft^2 °F	W/m^2 °C
Brick exterior wall, plaster interior, uninsulated	0.45	2.55
Frame exterior wall, plaster interior, uninsulated	0.25	1.42
With rock-wool insulation	0.07	0.4
Plate-glass window	1.10	6.2
Double plate-glass window	0.40	2.3
Steam condenser	200–1000	1100–5600
Feedwater heater	200–1500	1100–8500
Freon-12 condenser with water coolant	50–150	280–850
Water-to-water heat exchanger	150–300	850–1700
Finned-tube heat exchanger, water in tubes, air across tubes	5–10	25–55
Water-to-oil heat exchanger	20–60	110–350
Steam to light fuel oil	30–60	170–340
Steam to heavy fuel oil	10–30	56–170
Steam to kerosene or gasoline	50–200	280–1140
Finned-tube heat exchanger, steam in tubes, air over tubes	5–50	28–280
Ammonia condenser, water in tubes	150–250	850–1400
Alcohol condenser, water in tubes	45–120	255–680
Gas-to-gas heat exchanger	2–8	10–40

$$\text{Number of tubes} = \frac{278}{0.303} = 918$$

- Tube-bundle diameter

$$D_b = D_o(N_t/K_1)^{\frac{1}{n_1}} \tag{1.30}$$

D_b diameter of tube-bundle;
N_t number of tubes;
D_o outer diameter of the tube;
K_1 and n_1 constants determined from Table 1.7 based on passes number and tube arrangement.

$$\text{Eq. (1.30)}: \; D_b = 20\left(\frac{918}{0.249}\right)^{\frac{1}{2.207}} = 826 \text{ mm}$$

- An empty space is considered between shell wall and tube-bundle. This empty space depends on heat exchanger type and manufacturing tolerances. Figure 1.21 shows the required empty space for different heat exchanger types according to tube-bundle diameter. If the split-ring floating head heat exchanger is selected; from Fig. 1.21, it is clear that the required clearance is equal to 68 mm. As a result, shell diameter (D_s) is:

$$D_s = 826 + 68 = 894 \text{ mm}$$

- The closest pipe manufacturing standard to the above number is 863 mm (34 in.) or 914.4 mm (36 in.).

Table 1.7 Constants used in Eq. (1.30) [93]

Triangular pitch $P_t = 1.25\ d_o$					
No passes	1	2	4	6	8
K_1	0.319	0.249	0.175	0.0743	0.0365
n_1	2.142	2.207	2.285	2.499	2.675
Square pitch $P_t = 1.25\ d_o$					
No passes	1	2	4	6	8
K_1	0.215	0.153	0.158	0.0420	0.0331
n_1	2.207	2.291	2.263	2.617	2.643

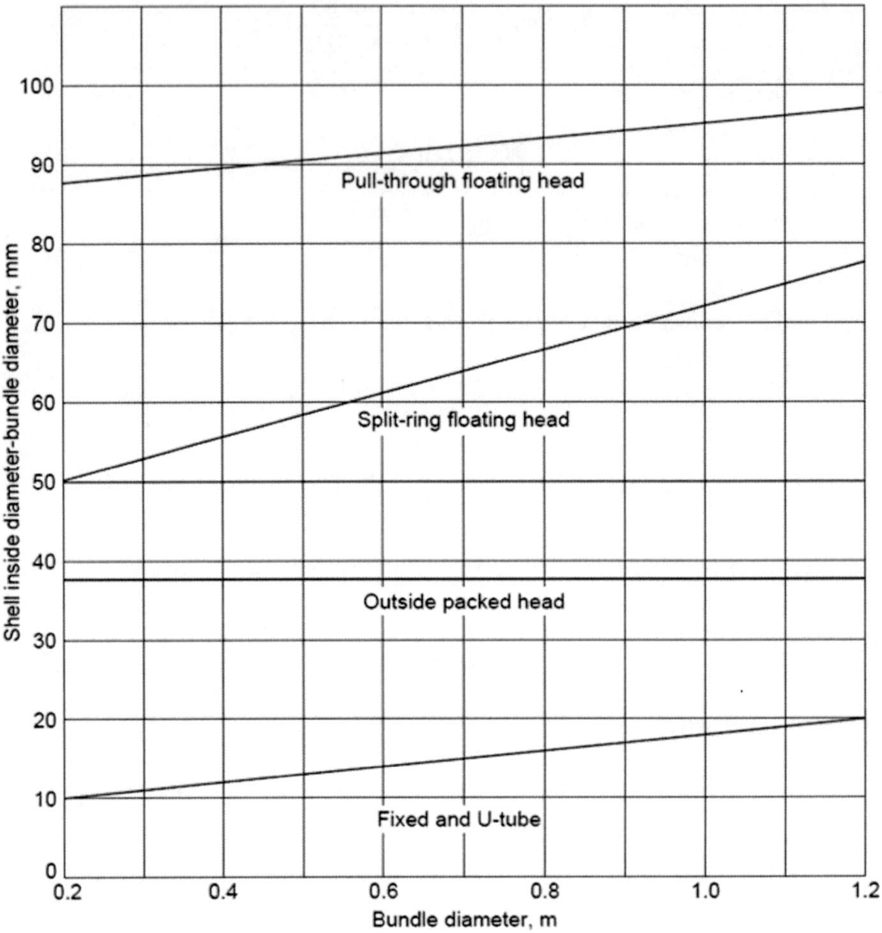

Fig. 1.21 Shell-bundle clearance [93]

- Next, the fouling factor is determined according to the fluids in service (R_f = 6000 W/m^2 °C). The important point here is that in the laboratory work, it was found that choosing the fouling factor as a fixed number can lead to problems such as the impossibility of predicting the time when the heat exchanger will be taken out of service for fouling removal.

Heat Transfer Coefficient of Fluid on the Tube Side

- Bulk fluid average temperature of tube side:

$$T_{avg} = \frac{40 + 25}{2} = 33\,^{\circ}\mathrm{C}$$

- Internal cross section area for a tube:

$$a = \frac{\pi}{4} D^2 = \frac{\pi}{4} \times 16^2 = 201\ \mathrm{mm}^2$$

- Due to the existence of two tube passes, to find the number of tubes in each pass, we divide the number of tubes by two:

$$tube\ number\ in\ each\ pass : \frac{918}{2} = 459$$

- Now, by multiplying the number of tubes in each pass by the cross-sectional area of one tube, one can find the cross-sectional area of all tubes in each pass:

$$all\ tube\ cross{-}sectional\ area\ in\ each\ pass = 459 \times 201 \times 10^{-6} = 0.092\ \mathrm{m}^2$$

- The water mass velocity is obtained by dividing the water mass flow by this cross section:
- water mass velocity:

$$G_t = \frac{\dot{m}}{A} = \frac{68.9}{0.092} = 749\ \frac{\mathrm{kg}}{\mathrm{s\,m}^2}$$

$$@33\,^{\circ}\mathrm{C} \rightarrow \rho = 995\ \mathrm{kg/m}^3$$

- water linear velocity:

$$u_t = \frac{G_t}{\rho} = \frac{749\ \mathrm{kg/s\,m}^2}{995\ \mathrm{kg/m}^3} = 0.75\,\mathrm{m/s}$$

- From Eq. (1.31), the heat transfer coefficient for water inside the tube can be obtained (this relation was developed by Eagle and Ferguson [94] specifically for water).

$$\begin{aligned} h_i &= \frac{4200\left(1.35 + 0.02\ T_{avg}\right)u_t^{0.8}}{D_i^{0.2}} \\ &= \frac{4200(1.35 + 0.02 \times 33)0.75^{0.8}}{16^{0.2}} = 3852\ \frac{\mathrm{W}}{\mathrm{m}^2\,^{\circ}\mathrm{C}} \end{aligned} \tag{1.31}$$

- This coefficient can also be obtained using Eq. (1.32) as follows:

$$Nu = \frac{h_i D_i}{k_f} = j_h Re Pr^{0.33}\left(\frac{\mu}{\mu_w}\right)^{0.14} \quad (1.32)$$

$$\mu = 0.8\,\frac{\text{mNs}}{\text{m}^2};\; k_f = 0.59\,\frac{\text{W}}{\text{m}\,^\circ\text{C}};$$

$$Re = \frac{\rho u_t D_i}{\mu} = \frac{995 \times 0.75 \times 16 \times 10^{-3}}{0.8 \times 10^{-3}} = 14{,}925;$$

$$Pr = \frac{C_p \mu}{k_f} = \frac{4.2 \times 10^3 \times 0.8 \times 10^{-3}}{0.59} = 5.7$$

Ignoring the ratio of (μ/μ_w) and calculating the ratio of L/di, j_h is calculated according to Fig. 1.22.

$$L/D_i = \frac{4.83 \times 10^3}{16} = 302 \rightarrow j_h = 3.9 \times 10^{-3}$$

$$\rightarrow \text{Eq. (1.32)}:\; h_i = \frac{0.59}{16 \times 10^{-3}} \times 3.9 \times 10^{-3} \times 14{,}925 \times 5.7^{0.33} = 3812\,\frac{\text{W}}{\text{m}^2\,^\circ\text{C}}$$

to calculate the viscosity correction factor, the wall temperature is needed, which can be estimated from Eq. (1.33):

$$h_i\left(t_{w,est} - T_{b,t}\right) = U\left(T_{b,s} - T_{b,t}\right) \quad (1.33)$$

where $t_{w,est}$ is estimated wall temperature, $T_{b,t}$ and $T_{b,s}$ are bulk fluid average temperature of tube side and shell side respectively.

Heat Transfer Coefficient of Fluid on the Shell Side

- Usually, the distance between the baffles is considered to be between 0.2 and 1 times the shell diameter.
- Baffle spacing:

$$l_B = \frac{D_s}{5} = \frac{894}{5} = 178\text{ mm}$$

- Distance between two adjacent tubes (pitch to pitch):

$$tube\ pitch = 1.25 \times D_o = 1.25 \times 20 = 25\text{ mm}$$

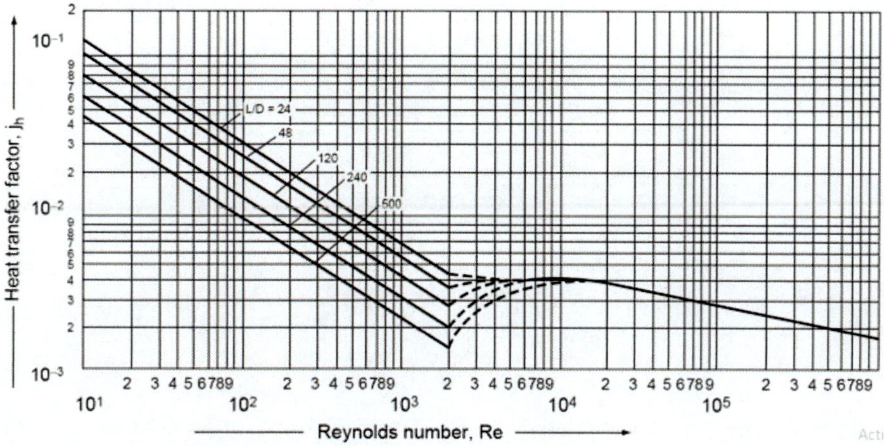

Fig. 1.22 Tube-side heat-transfer factor [93]

- Cross-sectional area of shell side flow pass:

$$A_s = \frac{(p_t - D_o)D_s l_B}{p_t} \tag{1.34}$$

where p_t: tube pitch; D_o: tubes outer diameter; D_s: shell inner diameter and l_B: baffle spacing.

$$l_B = \frac{(25 - 20) \times 894 \times 178 \times 10^{-6}}{25} = 0.032 \text{ m}^2$$

- Mass velocity:

$$G_s = \frac{\dot{m}}{A} = \frac{100{,}000}{3600} \times \frac{1}{0.032} = 868 \text{ kg/s m}^2$$

- Equivalent diameter for triangular arrangement:

$$D_e = \frac{1.1}{D_o}\left(p_t^2 - 0.917 D_o^2\right) = \frac{1.1}{20}\left(25^2 - 0.917 \times 20^2\right) = 14.4 \text{ mm} \tag{1.35}$$

- Bulk fluid average temperature of shell side:

$$T_{avg} = \frac{95 + 45}{2} = 68\,^\circ\text{C}$$

$$@68\,^\circ\text{C}:\ \rho = 750\,\text{kg/m}^3;\ \mu = 0.34\,\frac{\text{mNs}}{\text{m}^2};$$

$$C_p = 2.84\ \text{kJ/kg}\,^\circ\text{C};\ k_f = 0.19\,\frac{\text{W}}{\text{m}\,^\circ\text{C}}$$

$$Re = \frac{G_s D_e}{\mu} = \frac{868 \times 14.4 \times 10^{-3}}{0.34 \times 10^{-3}} = 36{,}762;$$

$$Pr = \frac{C_p \mu}{k_f} = \frac{2.84 \times 10^3 \times 0.34 \times 10^{-3}}{0.19} = 5.1$$

- If 25% baffle cuts are used, using Fig. 1.23:

$$j_h = 3.3 \times 10^{-3}$$

- By removing the term related to viscosity correction, it is concluded from Eq. (1.36):

$$Nu = \frac{h_s D_e}{k_f} = j_h Re Pr^{0.33} \left(\frac{\mu}{\mu_w}\right)^{0.14} \tag{1.36}$$

$$h_s = \frac{0.19}{14.4 \times 10^{-3}} \times 3.3 \times 10^{-3} \times 36{,}762 \times 5.1^{0.33} = 2740\ \text{W/m}^2\,^\circ\text{C}$$

- Now, can estimate the wall temperature to check the effect of (μ/μ_w) factor.
- Average temperature difference for all resistances:

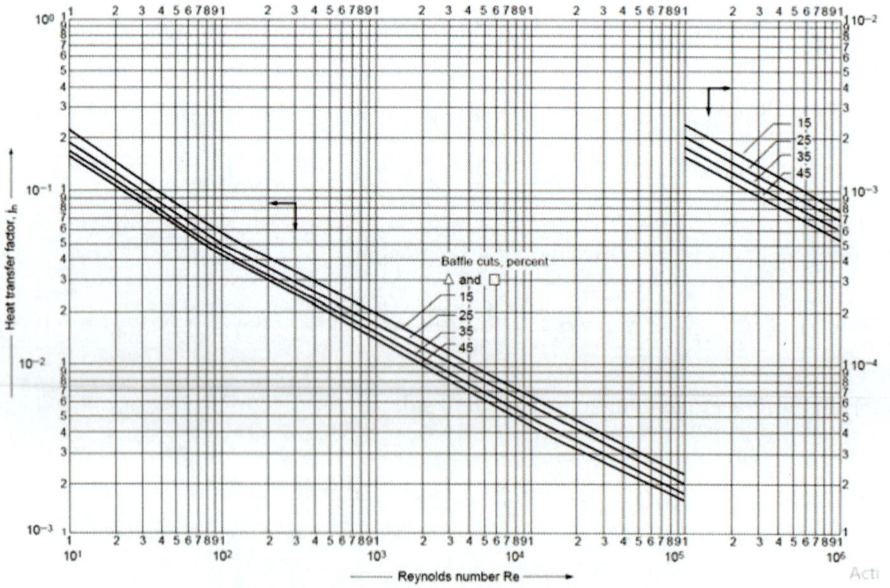

Fig. 1.23 Shell-side heat-transfer factors, segmental baffles [93]

$$temperature\ difference: 68 - 33 = 35\,^{\circ}\mathrm{C}$$

- Only for the resistance caused by the methanol film:

$$across\ methanol\ film = \frac{U}{h_o} \times \Delta T = \frac{600}{2740} \times 35 = 8\,^{\circ}\mathrm{C}$$

- Average wall temperature:

$$68 - 8 = 60\,^{\circ}\mathrm{C}$$

$$\mu_w = 0.37\,\frac{\mathrm{mNs}}{\mathrm{m}^2} \rightarrow \left(\frac{\mu}{\mu_w}\right)^{0.14} = 0.99$$

which shows that the viscosity correction factor is not important for fluids with low viscosity.

Overall Heat Transfer Coefficient

The thermal conductivity of cupro-nickel alloy is equal to 50 W/m °C, and the external dirty coefficients (heat transfer coefficient after deposit formation on the tube surface) from Table 1.8 are considered 5000 W/m^2 °C for methanol (light organic) and 3000 W/m^2 °C for water (sea water). Therefore, from Eq. (1.37) (which is presented in a different way in the previous sections), we will have:

$$\frac{1}{U} = \frac{1}{h_o} + \frac{1}{h_{od}} + \frac{D_o \ln(D_o/D_i)}{2k_w} + \frac{D_o}{D_i} \times \frac{1}{h_{id}} + \frac{D_o}{D_i} \times \frac{1}{h_i} \tag{1.37}$$

where

- U overall heat transfer coefficient; W/m^2 °C
- h_o heat transfer coefficient in the external fluid; W/m^2 °C
- h_i heat transfer coefficient in the internal fluid; W/m^2 °C
- h_{od} external dirty coefficient (heat transfer coefficient after creating fouling on the tube outer surface); W/m^2 °C
- h_{id} internal dirty coefficient (heat transfer coefficient after creating fouling on the tube inner surface); W/m^2 °C
- k_w thermal conductivity of the tube wall; W/m °C
- D_i tube inner diameter, m
- D_o tube outer diameter, m

$$\frac{1}{U} = \frac{1}{2740} + \frac{1}{5000} + \frac{20 \times 10^{-3} \ln(20/16)}{2 \times 50} + \frac{20}{16} \times \frac{1}{3000} + \frac{20}{16} \times \frac{1}{3812}$$
$$= 738\ \mathrm{W/m^2\,^{\circ}C}$$

Table 1.8 Fouling factors (coefficients)

Fluid type	Coefficient (W/m^2 °C)	Factor (resistance) (m^2 °C/W)
River water	3000–12,000	0.0003–0.0001
Sea water	1000–3000	0.001–0.0003
Cooling water (towers)	3000–6000	0.0003–0.00017
Towns' water (soft)	3000–5000	0.0003–0.0002
Towns' water (hard)	1000–2000	0.001–0.0005
Steam condensate	1500–5000	0.00067–0.0002
Steam (oil free)	4000–10,000	0.0025–0.0001
Steam (oil traces)	2000–5000	0.0005–0.0002
Refrigerated brine	3000–5000	0.0003–0.0002
Air and industrial gases	5000–10,000	0.0002–0.0001
Flue gases	2000–5000	0.0005–0.0002
Organic vapors	5000	0.0002
Organic liquids	5000	0.0002
Light hydrocarbons	5000	0.0002
Heavy hydrocarbons	2000	0.0005
Boiling organics	2500	0.0004
Condensing organics	5000	0.0002
Heat transfer fluids	5000	0.0002
Aqueous salt solutions	3000–5000	0.0003–0.0002

The calculated value is higher than the value assumed at the beginning of the calculations.

Calculation of Pressure Drop

- Tube side pressure drop

From Fig. 1.24 for Re = 14,925, friction factor is obtained:

$$j_f = 4.3 \times 10^{-3}$$

If the viscosity correction factor is ignored, the pressure drop from Eq. (1.38) will be equal to:

$$\Delta P_t = N_p\left[8 \times j_f \times (L/D_i)\left(\frac{\mu}{\mu_w}\right)^{-m} + 2.5\right]\frac{\rho u_t^2}{2} \qquad (1.38)$$

where

ΔP_t tube side pressure drop, N/m^2 (Pa)
N_p number of tube passes

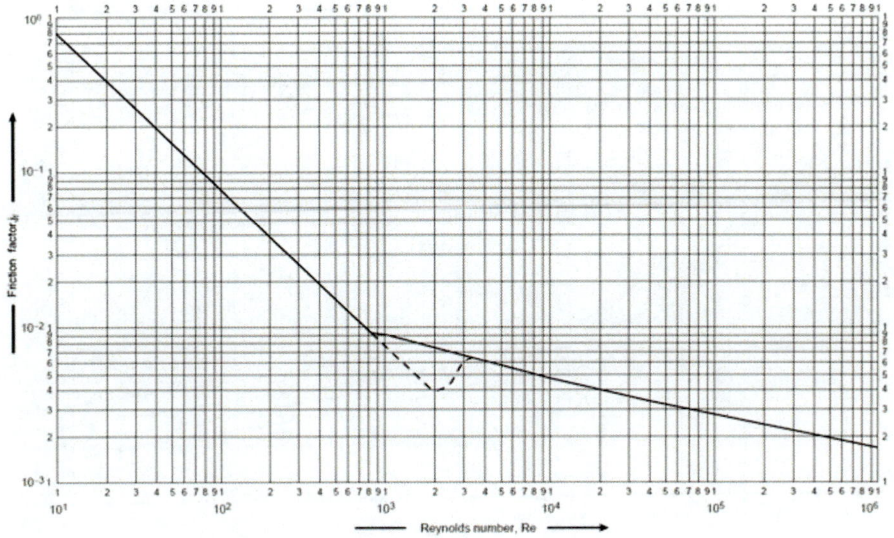

Fig. 1.24 Tube-side friction factors [93]

u_t tube side velocity, m/s
L tube length, m

$$\Delta P = 2 \times \left[8 \times 4.3 \times 10^{-3}\left(4.83 \times \frac{10^3}{16}\right) + 2.5\right]\frac{995 \times 0.75^2}{2}$$
$$= 7211\ \frac{N}{m^2} = 7.2\ kPa$$

which is low; for correction, increasing the number of tube passes can be considered.

- Shell side pressure drop
- Linear velocity

$$u_s = \frac{G_s}{\rho} = \frac{868}{750} = 1.16\ \frac{m}{s}$$

Using Fig. 1.25 and Re = 36,762:

$$j_f = 4 \times 10^{-2}$$

If the viscosity correction factor is ignored, the pressure drop from Eq. (1.39) will be equal to:

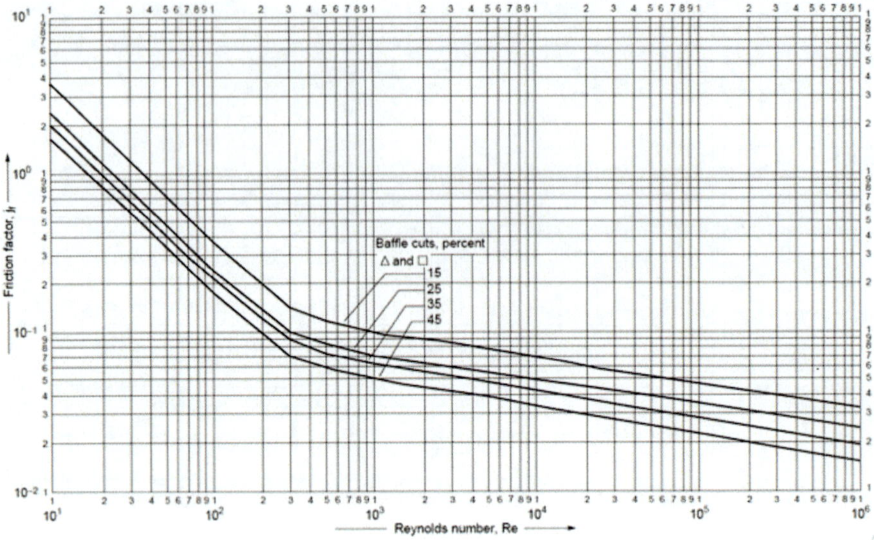

Fig. 1.25 Shell-side friction factors, segmental baffles [93]

$$\Delta p_s = 8 \times j_f \left[(D_s/D_e)(L/l_B) \frac{\rho u_s^2}{2} \left(\frac{\mu}{\mu_w} \right)^{-0.14} \right] \quad (1.39)$$

where

L tube length, m
l_B baffle spacing, m
(L/l_B) the number of times the flow passes through the tube-bundle

$$\Delta p_s = 8 \times 4 \times 10^{-2} \left[\frac{894}{14.4} \right] \left[\frac{4.83 \times 10^3}{178} \right] \frac{750 \times 1.16^2}{2} = 272{,}019 \frac{N}{m^2} = 272\ \text{kPa}$$

This amount is very high. This pressure drop can be reduced by increasing the distance of the tubes. Doubling the tube spacing, halves the shell-side fluid velocity, which leads to a reduction in pressure drop by a factor of about $(1/2)^2$ in this case:

$$\Delta p_s = \frac{272}{4} = 68\ \text{kPa}$$

This amount of pressure drop is acceptable. By applying these changes, the shell side heat transfer coefficient will change by a factor of $(1/2)^{0.8}$ (because $h_o \propto Re^{0.8} \propto u_s^{0.8}$). So:

$$h_o = 2740 \times (1.2)^{0.8} = 1573\ \text{W/m}^2\,^\circ\text{C}$$

This change reduces the overall heat transfer coefficient by 615 W/m^2 °C, which is still more than the assumed limit, so the calculations must be repeated.

References

1. Vosough A, Assari MR, Peyghambarzadeh SM, Azizi S. Influence of fluid flow rate on the fouling resistance of calcium sulfate aqueous solution in subcooled flow boiling condition. Int J Therm Sci 2020; 154: 106397.
2. Yang D, Liu J, Xiaoxue E, Jiang L. Experimental study of composition and influence factors on fouling of stainless steel and copper in seawater. Ann Nucl Energy 2016; 94: 767–772.
3. Wiechers HNS, Sturrock P, Marias GVR. Calcium carbonate crystallization kinetics. Water Res 1975; 9: 835–845.
4. Maddahi MH, Hatamipour MS, Jamialahmadi M. A model for the prediction of thermal resistance of calcium sulfate crystallization fouling in a liquid-solid fluidized bed heat exchanger with cylindrical particles. Int J Therm Sci 2019; 145: 106017.
5. Mwaba MG, Golriz MR, Gu J. A semi-empirical correlation for crystallization fouling on heat exchange surfaces. Appl Therm Eng 2006; 26: 440–447.
6. Crittenden BD, Yang M, Dong L, Hanson R, Jones J, Kundu K, Harris J, Klochok O, Arsenyev O, Kapustenko P. Crystallization fouling with enhanced heat transfer surfaces. Heat Transfer Eng 2015; 36: 741–749.
7. Hoang TA, Ang M, Rohl AL. Effects of process parameters on gypsum scale formation in pipes. Chem Eng Technol; 34(6): 1003–1009.
8. Mwaba MG, Rindt CCM, Van Steenhoven AA, Vorstman MAG. Experimental investigation of $CaSO_4$ crystallization on a flat plate. Heat Transfer Eng 2006; 27(3): 42–54.
9. Geddert T, Bialuch I, Augustin W, Scholl S. Extending the induction period of crystallization fouling through surface coating. Heat Transfer Eng 2009; 30(10–11): 868–875.
10. Kazi SN, Teng KH, Zakari MS, Sadeghinezhad E, Bakar MA. Study of mineral fouling mitigation on heat exchanger surface. Desalination 2015; 367: 248–254.
11. Zhao X, Chen XD. A critical review of basic crystallography to salt crystallization fouling in heat exchangers. Heat Transfer Eng 2013; 34(8–9): 719–732.
12. Geddert T, Augustin W, Scholl S. Induction time in crystallization fouling on heat transfer surfaces. Chem Eng Technol 2011; 34(8): 1303–1310.
13. Jamialahmadi M, Müller-Steinhagen H. A new model for the effect of calcium sulfate scale formation on pool boiling heat transfer. J Heat Trans 2004; 126: 507–517.
14. Babuška I, Silva RS, Actor J. Break-off model for $CaCO_3$ fouling in heat exchangers. Int J Heat Mass Tran 2018; 116: 104–114.
15. Abd-Elhady MS, Malayeri MR. Transition of convective heat transfer to subcooled flow boiling due to crystallization fouling. Appl Therm Eng 2016; 92: 122–129.
16. Riihimäki M, Pääkkönen TM, Guillot J, Lecoq E, Muurinen E, Simonson CJ, Keiski RL. Crystallization fouling on modified heat exchanger surfaces. In: Proceedings of the 12th International Conference on Heat Exchanger Fouling and Cleaning, Madrid, Spain, 2017, 155–162.
17. Kazi SN, Duffy GG, Chen XD. Fouling and fouling mitigation on heated metal surfaces. Desalination 2012; 288: 126–134.
18. Malayeri MR, Müller-Steinhagen H, Bartlett TH. Fouling of tube bundles under pool boiling conditions. Chem Eng Sci 2005; 60: 1503–1513.
19. Peyghambarzadeh SM, Bahrami N. Statistical analysis of calcium sulfate scaling under boiling heat transfer. Appl Therm Eng 2013; 53: 108–113.
20. Müller-Steinhagen H, Zhao Q, Helalizadeh A, Ren XG. The Effect of surface properties on $CaSO_4$ scale formation during convective heat transfer and subcooled flow boiling. Can J Chem Eng 2000; 78: 12–20.

21. Müller-Steinhagen H, Malayeri MR, Watkinson A. Heat exchanger fouling: environmental impacts. Heat Transfer Eng 2009; 30(10–11): 773–776.
22. O'Callaghan MG. Heat Exchanger Sourcebook. Washington DC: Hemisphere Publishing, 1986.
23. Shen C, Cirone C, Yang L, Jiang Y, Wang X. Characteristics of fouling development in shell-and-tube heat exchanger: Effects of velocity and installation location. Int J Heat and Mass Tran 2014; 77: 439–448.
24. Nesta J, Bennett CA. Fouling mitigation by design. In: Proceedings of 6th International Conference on Heat Exchanger Fouling and Cleaning - Challenges and Opportunities, Müller-Steinhagen H, Malayeri MR, Watkinson A, editors. Engineering Conferences International, Kloster Irsee, Germany, 2005, 342–347.
25. Bansal B, Chen XD, Müller-Steinhagen H. Analysis of 'classical' deposition rate law for crystallisation fouling. Chem Eng Process 2008; 47: 1201–1210.
26. Song KS, Lim J, Yun S, Kim D, Kim Y. Composite fouling characteristics of $CaCO_3$ and $CaSO_4$ in plate heat exchangers at various operating and geometric conditions. Int J Heat Mass Tran 2019; 136: 555–562.
27. Thackery FA. The cost of fouling in heat exchanger plant. Effluent Water Treat J 1980; 20(3).
28. Gilmour CH. No fooling-No fouling. Chem Eng Prog 1965; 61(7): 49–54.
29. Bansal B, Müller-Steinhagen H, Chen XD. Effect of suspended particles on crystallization fouling in plate heat exchangers. J Heat Trans 1997; 119(3): 568–574.
30. Förster M, Augustin W, Bohnet M, Influence of the adhesion force crystal/heat exchanger surface on fouling mitigation. Chem Eng Process 1999; 38: 449–461.
31. Heat Exchangers. (2013) Heat Exchanger Design Handbook, Second Edition (pp. 1–38): CRC Press.
32. Bansal, B., Chen, X.D., Müller-Steinhagen, H., 2003, Use of non-crystallising particles to mitigate crystallization fouling, International Communications in Heat and Mass Transfer, 30(5), 695–706.
33. Najibi, S.H., Müller-Steinhagen, H., Jamialahmadi, M., 1996, Boiling and nonboiling heat transfer to electrolyte solutions, Heat Transfer Engineering, 17(4) 46–63.
34. Müller-Steinhagen, H. M., 1993, Fouling: The ultimate challenge for heat exchanger design. The sixth International Symposium on Transport Phenomena in Thermal Engineering, Seoul, Korea.
35. Müller-Steinhagen, H., 1999, Cooling-water fouling in heat exchangers, Advances in Heat Transfer, 33, 415–496.
36. Pritchard, A. M., 1987, The Economics of fouling, in fouling science and technology, ed. L. F. Melo, T. R. Bott and C. A. Bernardo, NATO AS1 Series E, Vol. 145. (Dordrecht, Boston, London: Kluwer Academic Publishers).
37. Müller-Steinhagen, H., Malayeri, M. R., Watkinson, A. P., 2007, Recent advances in heat exchanger fouling research, Heat Transfer Engineering, 28(3), 173–176.
38. Lee, E., Jeon, J., Kang, H., Kim, Y., 2014, Thermal resistance in corrugated plate heat exchangers under crystallization fouling of calcium sulfate ($CaSO_4$), International Journal of Heat and Mass Transfer, 78, 908–916.
39. Müller-Steinhagen, H., Malayeri, M.R., Watkinson, A.P., 2011, Heat exchanger fouling: mitigation and cleaning strategies, Heat Transfer Engineering, 32(3–4), 189–196.
40. Dash, S., Rapoport, L., Varanasi, K. K., 2018, Crystallization-induced fouling during boiling: formation mechanisms to mitigation approaches, ACS Langmuir, 34(3), 782–788.
41. Shah RK, Sekulic D, 2003, Fundamentals of heat exchanger design, John Wiley & Sons, New York.
42. Bott, T. R., 1995, Fouling of Heat Exchangers, Elsevier, Amsterdam.
43. Müller-Steinhagen, H., (2000), Heat Exchanger Fouling: Mitigation and Cleaning Technologies, Publico publication, Germany.
44. Najibi, S.H., 1996, Heat transfer and heat transfer fouling during subcooled boiling for Electrolyte Solutions, Ph.D. thesis, Department of Chemical and Process Engineering, University of Surrey.

45. Malayeri, M. R., Müller-Steinhagen, H., 2007, Initiation of $CaSO_4$ scale formation on heat transfer surface under pool boiling conditions, Heat Transfer Engineering, 28(3), 240–247.
46. Fahiminia, F., Watkinson, A. P., Epstein, N., 2007, Calcium sulfate scaling delay times under sensible heating conditions, in Proceedings of the 6th International Conference on Fouling and Cleaning in Heat Exchangers, Tomar, Portugal, July 1–6, ECI Symposium Series, RP2, pp. 176–184.
47. Albert, F., Augustin, W., Scholl, S., 2009, Enhancement of heat transfer in crystallization fouling due to surface roughness, in Proceedings of Eurotherm Conference on Fouling and Cleaning in Heat Exchangers, Schladming, Austria, June 14–19, pp. 303–310.
48. Esawy M, Malayeri MR, Muller-Steinhagen H, (2010) Crystallization fouling of finned tubes during pool boiling effect of fin density. Heat Mass Transfer 46:1167–1176.
49. Jamialahmadi, M., Blochl, R., Müller-Steinhagen, H., 1989, Bubble Dynamics and Scale Formation During Boiling of Aqueous Calcium Sulphate Solutions, Chemical Engineering and Processing: Process Intensification, 26, pp. 15–26.
50. Peyghambarzadeh, S.M., Vatani, A., Jamialahmadi, M., 2013, Influences of bubble formation on different types of heat exchanger fouling, Applied Thermal Engineering, 50, 848–856.
51. Cai, Y., Liu, M., Hui, L., 2013, $CaCO_3$ Fouling on Microscale-Nanoscale Hydrophobic Titania-Fluoroalkylsilane Films in Pool Boiling, AIChE J, 59(7), 2662–2678.
52. Helalizadeh, A., Müller-Steinhagen, H., Jamialahmadi, M., 2000, Mixed salt crystallization fouling, Chem. Eng. Process, 39, 29–43.
53. Pääkkönen, T.M., Riihimäki, M., Simonson, C.J., Muurinen, E., Keiski, R.L., 2012, Crystallization fouling of $CaCO_3$ - Analysis of experimental thermal resistance and its uncertainty, International Journal of Heat and Mass Transfer, 55, 6927–6937.
54. Mullin, J. W., Crystallization, 3rd ed., Butterworth-Heinemann, Oxford, UK, 1993.
55. Esawy, M., 2011, Fouling of Structured Surfaces during Pool Boiling of Aqueous Solutions, Ph.D. thesis, Institute for Thermodynamics and Thermal Engineering, University of Stuttgart.
56. Varnaseri, M., Peyghambarzadeh, S.M. Interference effect of suspended particles on the crystallization fouling: A critical review. Heat Mass Transfer 59, 655–680 (2023).
57. Epstein, N. (1983). Thinking about Heat Transfer Fouling: A 5 $\times$ 5 Matrix. Heat Transfer Engineering, 4(1), 43–56.
58. Bohnet, M. (1987). Fouling of heat transfer surfaces. Chemical engineering & technology, 10(1), 113–125.
59. Watkinson, A. P. and Epstein, N., 1969, Gas Oil Fouling in a Sensible Heat Exchanger, Chem. Eng. Prog. Symp. Ser., 65 (92), 84–90.
60. Troup DH, Richardson JA, 1978, Scale nucleation on a heat transfer surface and its prevention, Chem Eng Commun, 2 (4–5), 167–180.
61. Baier, R. E., Early Events of Micro-Biofouling of All Heat Transfer Equipment, in Fouling of Heat Transfer Equipment, eds. E. F. C. Somerscales and J. G. Knudsen, pp. 293–304, Hemisphere, Washington, D.C., 1981.
62. Packter, A. (1968). The precipitation of sparingly soluble alkaline-earth metal and lead salts: nucleation and growth orders during the induction period. Journal of the Chemical Society A: Inorganic, Physical, Theoretical, 859–862.
63. Uyak, V., Akdagli, M., Cakmakci, M., & Koyuncu, I. (2014). Natural organic matter removal and fouling in a low pressure hybrid membrane systems. The Scientific World Journal, 2014.
64. Awad, M.M., Gad, H.E., Yousef, A., 2009, Effect of Surface Temperature on Particulate and Crystallization Fouling, Thirteenth International Water Technology Conference, IWTC 13, Hurghada, Egypt, 1431–1442.
65. Najibi, S.H., Müller-Steinhagen, H., Jamialahmadi, M., 1997, Calcium Carbonate Scale Formation during Subcooled Flow Boiling, Journal of Heat Transfer, 119, 767–775.
66. Schlichting, H., 1968, Boundary Layer Theory, 6th ed., McGraw-Hill Book Co., Inc., New York, pp. 596–625.
67. Crittenden, B.D., Aldermann, N.J., 1988, Negative fouling resistance: Effect of surface roughness, Chemical Engineering Science, 43(4), 829–838.

68. Tai, Y. T., Chen, F. B., 1998, Polymorphism of $CaCO_3$ Precipitated in a Constant Composition Environment, AIChE Journal, 44(8), pp. 1790–1798.
69. Morse, J. W., Arvidson, R. S., Luttge, A., 2007, Calcium Carbonate Formation and Dissolution, Chem. Rev., 107(2), pp. 342–381.
70. Kamhi, S. R., 1963, On the Structure of Vaterite $CaCO_3$, Acta Crystallographica, 16(16), pp. 770–772.
71. Wray, J. L., Daniels, F., 1957, Precipitation of Calcite and Aragonite, Journal of the American Chemical Society, 79(9), pp. 2031–2034.
72. Reeder, R. J., 1990, Carbonates, Mineralogy and Chemistry, Washington DC: Mineralogy Society of America.
73. Hasson, D., Zahavi, J., 1970, Mechanism of Calcium Sulfate Scale Deposition on Heat Transfer Surfaces, I&EC Fundamentals, 9(1), pp. 1–10.
74. Lager, G. A., 1984, A Crystallographic Study of the Low-Temperature Dehydration Products of Gypsum, $CaSO_4 \cdot 0.5H_2O$, and $CaSO_4$, American Mineralogist, 69, pp. 910–918.
75. Sheikholeslami, R., Ng, M., 2001, Calcium Sulfate Precipitation in the Presence of Nondominant Calcium Carbonate: Thermodynamics and Kinetics, Industrial & Engineering Chemistry Research, 40(16), 3570–3578.
76. Hasson, D., 1981, Precipitation Fouling-A Review, in Fouling of Heat Transfer Equipment, eds. E.F.C. Somerscales, and J. G. Knudsen, Hemisphere, New York, NY, pp. 527–568.
77. Hasson, D., 1999, Progress in Precipitation Fouling Research, A Review, Proc. Eng. Found. Int. Conference, Understanding Heat Exchanger Fouling and its Mitigation, New York, NY, pp. 67–89.
78. Höfling V., Augustin, W., Bohnet, M., 2004, Crystallization Fouling of the Aqueous Two-component System $CaSO_4/CaCO_3$, International Journal of Transport Phenomena, 6, pp. 99–109.
79. Hasson, D., Karmon, N., 1984, Novel Process for Lining Water Mains by Controlled Calcite Deposition, Corrosion Prevention and Control, 31, pp. 9–17.
80. Najibi, S.H., Müller-Steinhagen, H., Jamialahmadi, M., 1997, Calcium sulphate scale formation during subcooled flow boiling, Chemical Engineering Science, 52(8), 1265–1284.
81. J. P. Holman, Heat Transfer, 10th ed., McGraw-Hill series in mechanical engineering, 2010.
82. Sadik Kakaç, Hongtan Liu, Anchasa Pramuanjaroenkij, Heat Exchangers, Selection, Rating, and Thermal Design, Third Edition, CRC Press, Taylor & Francis Group, New York.
83. Varnaseri, M., Peyghambarzadeh, S.M. Particulate fouling during boiling heat transfer and crystallization of $CaCO_3$ aqueous solutions. Heat Mass Transfer 59, 1477–1505 (2023).
84. V Nikkhah, MM Sarafraz, F Hormozi, SM Peyghambarzadeh, Particulate fouling of CuO–water nanofluid at isothermal diffusive condition inside the conventional heat exchanger-experimental and modeling, Experimental Thermal and Fluid Science, 60, 2015, 83–95.
85. On the fouling formation of functionalized and non-functionalized carbon nanotube nano-fluids under pool boiling condition, MM Sarafraz, F Hormozi, M Silakhori, SM Peyghambarzadeh, Applied Thermal Engineering, 95, 2016, 433–444.
86. SM Peyghambarzadeh, A Vatani, M Jamialahmadi, Application of asymptotic model for the prediction of fouling rate of calcium sulfate under subcooled flow boiling, Applied Thermal Engineering, 39, 2012, 105–113.
87. SM Peyghambarzadeh, A Vatani, M Jamialahmadi, Experimental study of micro-particle fouling under forced convective heat transfer, Brazilian Journal of Chemical Engineering, 29(4):713–724, 2012.
88. A Vosough, SM Peyghambarzadeh, MR Assari, Influence of thermal shock on the mitigation of calcium sulfate crystallization fouling under subcooled flow boiling condition, Applied Thermal Engineering, 164, 2020, 114434.
89. H Hallaji, SM Peyghambarzadeh, MR Bohloul, S Azizi; The optimum conditions for calcium sulfate fouling rate under subcooled flow boiling using Taguchi statistical method, Int J Heat Mass Trans, 204, 2023, 123859.
90. S Kamalifar, SM Peyghambarzadeh, S Azizi, F Jamali-Sheini, Experimental study on crude oil fouling in preheat exchangers at different operating conditions, Thermal Science and Engineering Progress, 39, 2023, 101742.

91. S Rahimi, SM Peyghambarzadeh, H KazemiEsfeh, S Azizi, MR Malayeri, Air rumbling for online fouling mitigation of calcium carbonate aqueous solution under subcooled flow boiling heat transfer, Applied Thermal Engineering, 225, 2023, 120159.
92. S Rahimi, SM Peyghambarzadeh, H KazemiEsfeh, S Azizi, MR Malayeri, Experimental investigation of temperature and hydrodynamics on $CaCO_3$ fouling during convective heat transfer and subcooled flow boiling, Applied Thermal Engineering, 220, 2023, 119698.
93. Gavin P. Towler, R.K. Sinnott, Chemical Engineering Design, 5th ed., Amsterdam: Elsevier/ Butterworth-Heinemann, 2008.
94. Eagle A, Ferguson, RM, 1930, On the coefficient of heat transfer from the internal surfaces of tube walls. Proc Roy Soc. A. 127, 540.

Chapter 2
Induction Period in Crystallization Fouling

Abstract A period of the fouling process in which there is no visible deposition and the fouling resistance is not yet positive is usually called the induction period. Depending on the heat transfer and operational conditions, the induction period may be occurred or not. Applying methods that lead to an increase in the induction period has recently been considered as modern fouling mitigation strategies. Contrary to the relatively large studies that have been conducted to identify and investigate the factors affecting the fouling growth period, operational parameters effect on induction period are not fully understood and quantified so far. If there is information about the cause of induction period and the factors affecting it, the induction period can be increased and fouling problems in heat exchangers can be greatly reduced or postponed. This study purpose is to investigate and summarize the operational parameters (e.g., fluid velocity, surface and bulk temperature, salt concentration and heat flux) affecting the induction period of calcium salts (e.g., $CaSO_4$, $CaCO_3$ and mixed of them) to identify the areas that have received less attention in the last 50 years.

Keywords Crystallization fouling · Calcium carbonate · Calcium sulfate · Induction period · Operational parameters

2.1 Introduction

If the behavior of fouling resistance (R_f) is plotted versus time for a crystallization fouling, three main zones will be identified. The experimental results of Ritter [1] have shown that unlike the asymptotic cooling water fouling curve, the curve with an induction period which there is no visible deposition and then constant fouling rate are characteristic of crystallization fouling. Figure 2.1 shows a characteristic fouling curve for a $CaCO_3$ solution [2].

In Fig. 2.1, the first zone is where the resistance becomes zero and then negative. Several authors have reported negative fouling resistances [3]. At first, no deposition is observed for some times and fouling resistance remained unchanged during this period (fouling resistance remains zero). During the crystallization fouling, a series of

M. Varnaseri and S. M. Peyghambarzadeh, *Scale Formation in Heat Exchangers*,
https://doi.org/10.1007/978-3-031-52704-3_2

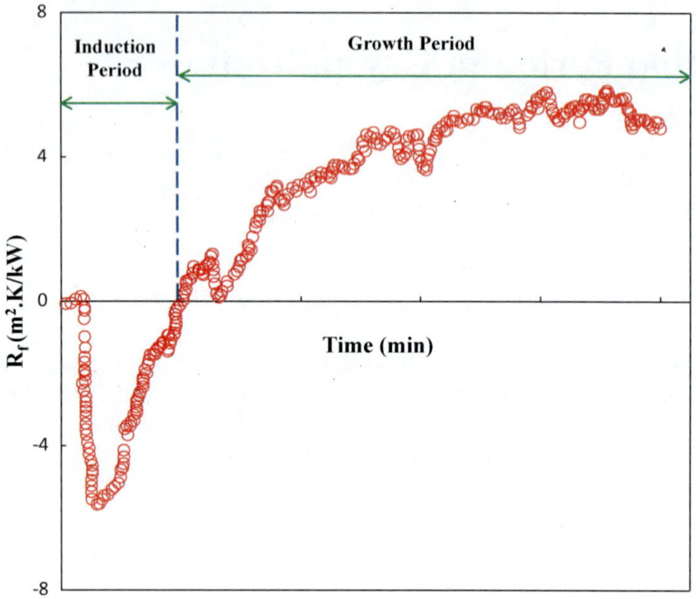

Fig. 2.1 Different time zone of crystallization fouling (adapted from [2])

consecutive events usually occur, and the induction period (initiation) is the first stage [4]. Penetration of the initial layer of grown fouling to the viscous sublayer leads to an increase in the surface roughness and as a result, increases the flow turbulence. As a result, the convective heat transfer increases, and fouling resistance becomes negative [5]. The reason for positive fouling resistance with the formation of more deposition is the predominance of the increase in thermal resistance over the turbulence created by the initial deposition. The time interval from the start until the fouling resistance becomes positive, after being negative for some times, is called induction time [6]. Then, there is a zone where the fouling resistance is continuously increasing, which is called the fouling growth zone. In this time interval, the growth of crystals increases and a layer of fouling forms on the surface. Wang et al. [7] defined the induction time as the time that elapses between the moment the supersaturation condition is achieved and the moment crystals are detected in the solution. Chuanfang [8] stated that when a layer with a very small thickness is formed on the heat transfer surface from the joining of crystals as the induction period. The lack of a significant decrease in heat transfer despite the continued growth of crystals is the description given by Kim et al. [9] as the induction period. The results of the studies by Albert et al. [10], and Abd-Elhady and Malayeri [11] include two significant points in the context of the induction period, one is the absence of significant fouling formation in it and the other is the end of the period with positive fouling resistance. Yang et al. [12] have stated that the extent of induction period may vary from a few minutes to many days. Peyghambarzadeh et al. [13] stated that the lack of continuity of the nucleation sites on the heat transfer surface during the initiation period leads to surface roughness and in the turbulent flow condition leads to an increase in heat transfer. Geddert et al. [5, 14] and Fahiminia et al. [15] categorized the factors influencing the induction period. They have divided these factors into two categories: (i) Process parameters, (ii) Interface parameters.

2.1.1 Process Parameters Affecting the Induction Period

Process parameters including working fluid, bulk and surfaces temperatures, supersaturation, pH, velocity and flow regime, and additives. Mwaba et al. [16] in a plate heat exchanger, by investigating the effect of flow conditions on $CaSO_4$ fouling, have stated that the flow velocity has a direct relationship with the induction period length. Rizzo et al. [17] proved that fluid velocity has a significant effect on nucleation (initiation period) as long as the fouling process is controlled by mass transfer. Fahiminia et al. [18] have investigated the operational parameters effective on the initiation period of $CaSO_4$ fouling, such as fluid velocity and heat transfer surface temperature. They state that at fluid velocities above 0.5 m/s, the initiation period is independent from the fluid velocity. Yang et al. [19] stated that for crystallization fouling during the growth period, due to the increased adhesion of the fouling layer to the heat transfer surface, it is very difficult to remove the fouling by increasing the flow velocity, but it is possible during the induction period. Saleh et al. [20] reported that induction period shortens with increased surface and bulk temperature, but extends by increased flow velocity. Zhenhua et al. [21] stated that the induction period gets shorter when the fluid velocity decreases, hardness and alkalinity increase, and solution temperature and heat transfer surface temperature increase. This brief overview shows the significant effect of process parameters on the induction period length.

2.1.2 Interface Parameters Affecting the Induction Period

This group of parameters depends on the heat transfer surface and includes topography, surface energy, roughness, number of nucleation sites and fouling layer aging. Najibi et al. [22] showed that the entire fouling process is affected by the surface properties and is not specific to the initiation fouling period. Yang et al. [19, 23] by comparing low-energy surfaces and copper surfaces found that these low-energy surfaces increased the induction period of $CaCO_3$ fouling and decreased the fouling rate. In addition, Herz et al. [24] showed that with increasing surface roughness, $CaSO_4$ fouling resistance increases and its induction period length decreases significantly. Their results recently confirmed by Stärk et al. [25]. Malayeri et al. [26] have introduced a modified surface of nano, which should be considered more because it has a much longer induction period and, on the other hand, the fouling rate are much lower than the other investigated surfaces. Kazi et al. [3] proved that some polymers are highly efficient in preventing the nucleation and crystallization of many soluble inorganic salts and prolong the induction time of calcium sulfate dehydrate crystallization. By introducing a two-dimensional parameter in the field of surface texture, Förster and Bohnet [27] have shown that the induction period is directly related to the mentioned parameter. Therefore, it can be concluded that the length of the induction period depends, among other parameters, on the energetic characteristics and topography of the heat transfer surface [28].

2.1.3 Anti-fouling Strategies Related to Induction Period

Increasing the length of the induction period has become the basis of new fouling reduction strategies, and these strategies include (1) increasing the duration of the induction period by reducing the adhesion of the fouling crystals to the heat transfer surface by changing the geometry and surface energy, (2) minimizing the fouling by selecting and adjusting the flow hydrodynamic conditions [29]. In addition, Müller-Steinhagen et al. [30] have classified many methods for online and offline fouling mitigation in heat exchangers. Their suggestion is to prevent fouling in the design stage, then use online methods to fouling mitigation, and the best strategy is actually a combination of these two. Some of the strategies used for fouling mitigation in heat transfer equipment with the aim of influencing the induction period length are reviewed below.

Mayer et al. [31] stated that a prolongation of the induction phase may be a very effective way to mitigate fouling effects in industrial applications. Malayeri and Müller-Steinhagen [32] showed that choosing the right method to fouling mitigation and understanding the mechanism of fouling is significantly dependent on the primary crystal nuclei that appear on the heat transfer surface during the induction period. The use of projectiles is an effective way to reduce fouling in the growth period, but in the induction period, regardless of the hardness and shape of the projectiles, the result was the opposite [4, 33]. In addition, they observed that air injection is not a suitable method to reduce the crystallization fouling and has increased the induction period [4]. Al-Janabi et al. [34] proved that grooves structure on tubes has led to a change in the induction period length, so that mixed and longitudinal grooves have a shorter induction period than cross grooves. The results of Föste et al.'s [35] studies on fouling mitigation methods have shown that pulsed flow is suitable as a method to reduce fouling by increasing fluid forces, although this method has led to a reduction in the induction period. Another result of their study was the reduction of induction period length by increasing the tubes roughness. Al-Janabi et al. [36] showed that shot peened surfaces are characterized by shorter induction time than that of the untreated stainless-steel surface. In practice, this means that the crystal nucleation process is expected to be faster as surface roughness increases. In addition, understanding the mechanism governing heat transfer equipment and the conditions under which the equipment operates has a great impact on the finding a suitable anti-fouling strategy. To understand the effect of different heat transfer conditions, can refer to the studies of our group, which are mostly in subcooled flow boiling [37–46], forced convection [39–42, 45, 47, 48] and pool boiling [49–59] conditions (which are commonly encountered in industry).

In general, by reviewing the research done in the field of fouling induction period, the following results can be achieved, which in fact are the cause of the present study:

- Reviewing the crystallization fouling studies in the past few decades, it was found that most of these studies have focused on the interfacial parameters, and less attention have been paid to the process parameters.
- Independent data can rarely be found on the effect of process parameters on crystallization fouling induction period, so, in the present study the data of various papers on the fouling of calcium salts in the early times of the experiments have been used to investigate the effect of process parameters.
- The entire fouling process is affected by the induction period. Furthermore, to create an effective anti-fouling method, the effect of all parameters including process parameters must be fully investigated. Our research shows that there is no complete and comprehensive research in this field and the present study is unique in its kind.

Our goal is to create general ideas among crystallization fouling research and to identify the effects of various factors on this type of fouling. In addition, it is tried to show the strengths and weaknesses of the previous researches and the possibility of creating new research areas. In the present chapter, similar data from different authors have been selected to investigate the effect of different operating parameters on the induction period at the early times of fouling formation. The studies in which the effects of the parameters shown more clearly were selected to investigate the effect of the parameters.

All these parameters (operational and interfacial) influence on the induction time as well as on the crystal growth period. Due to the widespread parameters affecting fouling, in this paper, only operational parameters including fluid velocity, temperature of heat transfer surface, fluid bulk temperature, salt concentration and heat flux applied to the heat transfer surface are discussed. In this paper, the operational factors affecting the crystallization fouling of calcium salts including calcium sulfate, calcium carbonate and their mixture are investigated.

2.2 Effects of Flow Velocity

To investigate the flow velocity effect on the induction period, among many studies that have been performed, some important results were selected. The selection criteria for the studies presented are:

- Display the data in the form of fouling resistance (R_f) against time.
- Presentation of the effect of parameters on the fouling resistance clearly.
- The data have been widely cited by other researchers.

There are three main views of flow velocity effect on the induction period length. The first view is that after the appearance of the first nuclei, the increase in the velocity causes the surface to become hydrodynamically rough and the shear stress on the surface increases, resulting in increased heat and mass transfer. By increasing the heat transfer, the heat transfer surface temperature decreasing and as a result, crystallization fouling is delayed. Wang et al. [60] by examining the effect of flow velocity on the deposition of $CaCO_3$ in smooth tubes showed that the maximum heat transfer coefficient and the fouling resistance minimum increase by increasing flow velocity (in the period of negative fouling resistance). In addition, the length of negative fouling resistance increases with increasing flow velocity and vice versa. This phenomenon has been described in several studies by changing flow properties at heat transfer surface, including the study of Yang et al. [12]. Pääkkönen et al. [61] showed that the duration of the induction period is sensitive to changes in operating conditions. They stated that as the flow velocity decreases, the convective heat transfer decreases, the wall temperature and the laminar boundary layer thickness increase, as a result, the reaction rate constant and the consequent crystallization fouling resistance increase, therefore, the induction period decreases. Najibi [62] has stated that the characteristics of the fouling are affected by the laminar sub-layer formed near the heat transfer surface, whose thickness increases with the decrease in the flow velocity and has a supersaturation degree and temperature different from the bulk flow. Hence, the fouling resistance decreases due to the increased turbulence caused by the surface roughness by the fouling diffusion into the laminar sub-layer. On the other hand, by increasing the flow velocity augments this turbulence and further reduces the fouling resistance. Förster et al. [29] investigated the adhesion force impact between heat exchanger surfaces-crystal on fouling reduction. In terms of flow velocity effect on the induction period length, they found that modifying geometry of surface should be directed towards increasing flow velocity or shear stress. This will increase the heat transfer and decrease the surface temperature; as a result, crystallization fouling will occur with more delay. Similar results have been observed in other studies [4, 9, 16, 62–72].

In the next view, the flow velocity effect on the induction period length was reported to be negligible. Hasson and Zahavi [73] conducted a study on $CaSO_4$ salt solutions using statistical methods. Their results showed that Reynolds number (Re) and shear stress have no effect on the induction period except in cases of dependence on the mass transfer coefficient. Mwaba et al. [16], Xiaokai et al. [69] and Han et al. [74] obtained similar results.

The third view states that with increasing flow velocity, more foulant ions reach the surface and as a result, fouling resistance increases and the induction period decreases. From this group of researchers, we can mention Yang et al. [19] who studied the induction period related to $CaCO_3$ deposition on the heat transfer surface. They implied that at a constant surface temperature (76.5 °C), the induction period increases as the flow velocity decreases. It should be noted that in their experiments to keep the surface temperature constant during increasing flow velocity, the heat flux applied to the surface increased gradually. However, they stated that if the heat flux applied to the heat transfer surface were constant increasing the flow velocity would reduce the surface temperature. It reduces the fouling rate and increases the induction period. Fahiminia et al. [15, 18] presented a different view, by examining different operating conditions on $CaSO_4$ deposition delay time. They concluded that at a constant bulk concentration and surface temperature, increasing the flow velocity up to about 0.5 m/s, the deposition delay time decreases and then remain constant. The reason for this phenomenon is the dependence of mass transfer on flow velocity and the control of the nucleation process by mass transfer at low flow velocity, and then, a velocity-independent surface process controls the deposition. There are very rare studies on the induction period of the deposition of aqueous salt mixtures. Sheikholeslami [75], who was a pioneering researcher in the mixed salt crystallization fouling, stated that the induction period is specific for each salt, and for a single salt in the crystallization fouling, the induction period of $CaSO_4$ is much longer than that of $CaCO_3$. She also stated that for mixed solutions (including $CaCO_3$ and $CaSO_4$ salts), the formation of $CaCO_3$ on the surface acts as nuclei for the formation of $CaSO_4$, and thus, it can be said that the induction period of the mixed solutions follows the same time pattern of pure $CaCO_3$. These results have been confirmed by other studies [76, 77], which have investigated the effect of other soluble materials and crystal species on the induction period. Since the Gibbs free energy of formation for heterogeneous nucleation is less than that for homogeneous nucleation, $CaCO_3$ acts as the nuclei for $CaSO_4$ and, as a result, nuclei form earlier in the salt mixture. Similar results have been observed in the research results of Hoang et al. [63], Najibi et al. [22, 78], Helalizadeh et al. [79] and Hasson et al. [80]. The results of the effect of flow velocity on the induction period of some of these studies, in the early time of fouling formation, are presented below.

Figures 2.2, 2.3, 2.4 and 2.5 shows the flow velocity effect on induction period for single and mixed salts. Figure 2.2 shows the fouling resistance as a function of time at two different velocities. It is taken from the first 150 min of the experiment performed by Peyghambarzadeh et al. [43], which was performed on $CaSO_4$ under subcooled flow boiling conditions. Other parameters of this research are C_b: 1.5–2.2 g/l, T_w: 102–115 °C and q: 400 kW/m^2. According to Fig. 2.2, by increasing the flow velocity from 0.5 to 2 m/s; the induction period increased by about 50%.

Figure 2.3 shows the effect of flow velocity on $CaSO_4$ crystallization fouling on a flat plate at velocities of 0.3–1 m/s corresponding to Reynolds numbers 11,000–34,000. These data were obtained by Mwaba et al. [16] at forced convective heat transfer, and the other operating conditions of this research are $C_{b,\,init}$: 3 kg/m^3, T_b: 40 °C and T_w: 65–82 °C. From Fig. 2.3, the direct relationship between induction period and Reynolds number (flow velocity) can be clearly seen. With the Reynolds number decrease, two effective factors in fouling formation, i.e. boundary layer thickness and fluid shear force, increase and decrease respectively. Increasing the boundary layer thickness leads to the creation of more ions and thus increasing the embryo formation chance; on the other hand, reducing the shear force will lead to

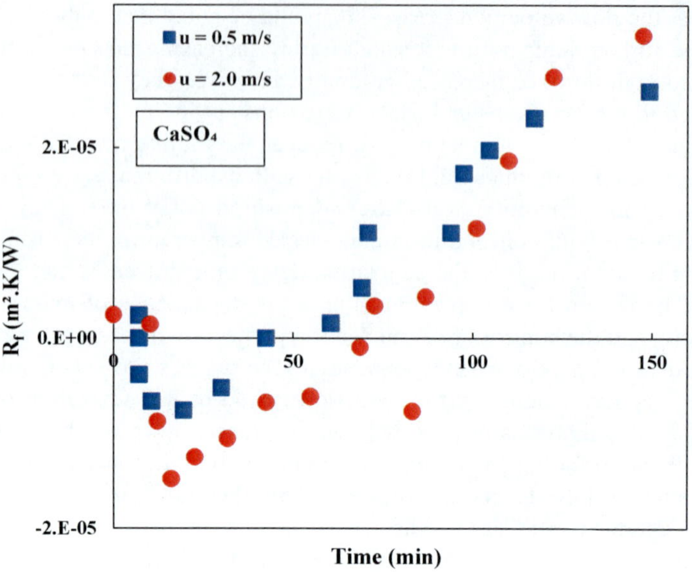

Fig. 2.2 Flow velocity effect on the induction period of $CaSO_4$ fouling (adapted from [43])

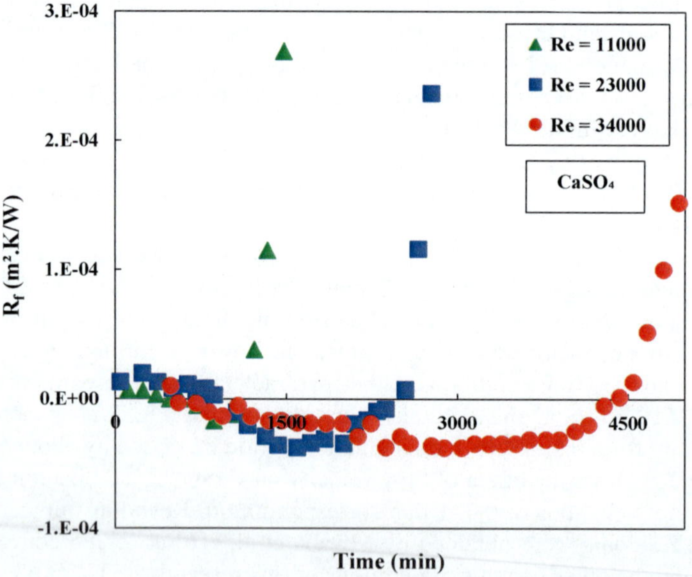

Fig. 2.3 Flow velocity effect on induction period of $CaSO_4$ crystallization fouling (adapted from [16])

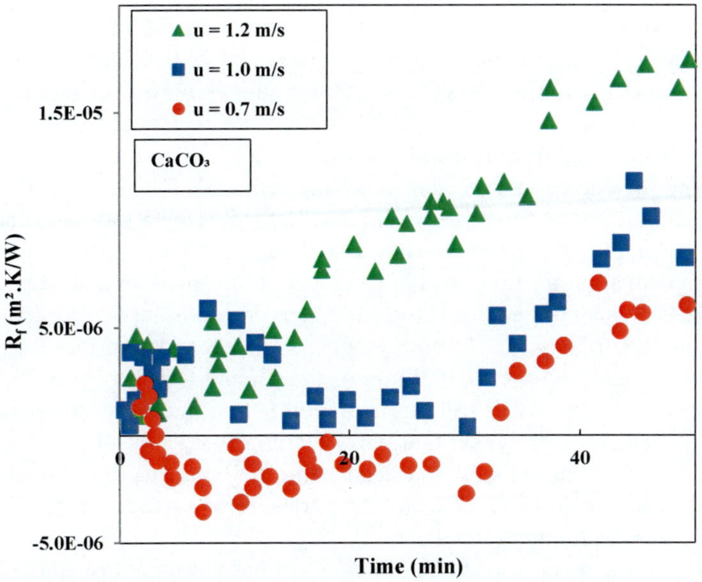

Fig. 2.4 Flow velocity effect on induction period of $CaCO_3$ fouling (adapted from [22])

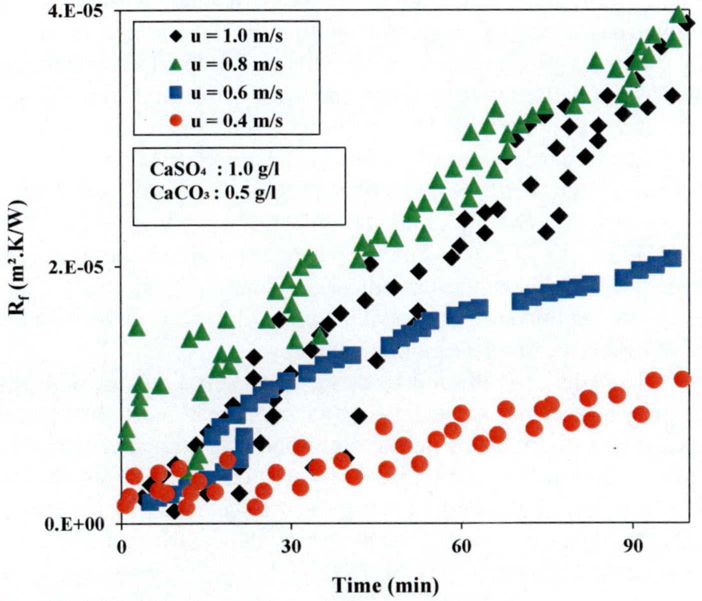

Fig. 2.5 Flow velocity effect on induction period of salt mixed precipitation (adapted from [79])

the particles staying longer on the surface and thus reducing the induction period. With increasing Reynolds from 11,000 to 23,000 and 34,000, the induction period increased 2 and 3.5 times, respectively. Other studies have also reported similar findings [29, 65, 78].

Figure 2.4 obtained from research by Najibi et al. [22] on the calcium carbonate ($CaCO_3$) fouling during subcooled flow boiling at the first 50 min of the experiment. Operating conditions of this research are C_b: 0.0035–0.0075 mol/l, T_b: 80 °C, T_s: 105–120 °C and u: 0.6–1.8 m/s. They stated that at all velocities studied in the initial period of fouling, the effects of nucleation are predominant. On the other hand, there is a general idea that if mass transfer, at a constant surface temperature, does not control fouling, the fouling process is independent of the flow velocity. However, by examining the first 50 min of the fouling test of Fig. 2.4, it is found that at a fluid velocity of 0.7 m/s, an induction period is almost detectable, but at higher flow velocities, the induction period is not practically observable. In fact, by increasing the fluid velocity and decreasing the boundary layer thickness, the controlling mechanism changes from mass transfer to surface chemical reaction, and it increases the fouling resistance.

Pääkkönen et al. [61] who studied $CaCO_3$ salt have also confirmed these results. Bansal and Müller-Steinhagen [81] have stated that the decrease in the wall temperature and viscose sub-layer thickness due to the increase in the turbulence in the flow velocity of more than 0.55 m/s will eventually lead to an increase in induction period length. Yang et al. [19] implied that when the heat flux is fixed, the surface initial temperature has decreased due to the increase in the flow velocity and due to the mutual effect of these two parameters, the induction period length has increased. These results have been confirmed by others researchers [21, 44].

Figure 2.5 reported by Helalizadeh et al. [79], showed the flow velocity effect on induction period of salt mixtures for a concentration of 1 g/l of $CaSO_4$ and 0.5 g/l of $CaCO_3$. Other operating conditions are u: 0.5–2 m/s, T_b: 50–90 °C, T_s: 103–117 °C, q: 100–400 kW/m^2. As shown in Fig. 2.5, there is no induction period for all four flow velocities studied. Under these conditions, calcium carbonate acts on the surface as nuclei to form calcium sulfate crystallization and in fact, mixed fouling follows the pattern of calcium carbonate fouling [78].

The flow velocity has two different roles: (1) Due to the direct relationship between the flow velocity and the mass transfer coefficient, increase in fouling will occur due to the increase in mass transfer; on the other hand, the transfer of more ions will intensify this process due to the increase in turbulence (mass transfer-controlled). (2) The flow velocity is proportional to the shear force, and with the increase of the flow velocity, the adhesion force of the deposit with heat transfer surface decreased (surface integration-controlled) [60, 61, 81].

Peyghambarzadeh et al. [37] provided an interesting comparison in terms of the effect of heat transfer mechanism as well as the effect of bubble formation on the induction time (first 100 min of experiment) of $CaSO_4$ crystallization fouling. Other operating conditions were considered the same. The result of this comparison is presented in Fig. 2.6. Fouling was formed faster on the surface under subcooled flow boiling conditions compared to the single-phase convective conditions. Figure 2.6

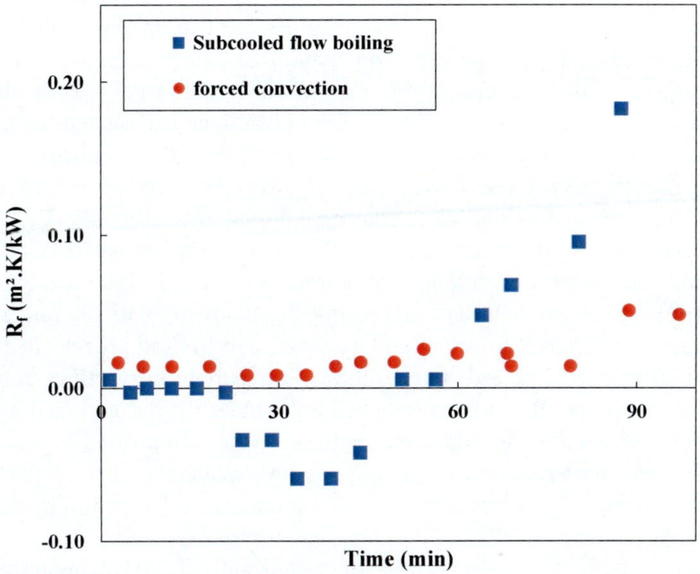

Fig. 2.6 Comparing the induction time of $CaSO_4$ crystallization fouling under subcooled flow boiling and single-phase forced convective heat transfer conditions (adapted from [37])

shows that bubble activity in subcooled flow boiling conditions compared to forced convective caused a severe increase in the fouling resistance and reduce the induction time. As mentioned earlier, when vapor bubbles grow on the heat transfer surface, the salt concentration below the vapor bubbles and near the heat transfer surface increases, and creates higher supersaturation in the boundary layer.

In general, it can be said that in the induction period, when the effects of additional disturbances that are directly related to velocity changes are greater than the effects of heat transfer area reduction, the increase in heat transfer coefficient overrides its reduction due to fouling, and fouling resistance becomes negative. On the other hand, when the effect of reduction in the surface area prevails, the fouling resistance increases from a minimum value (a negative value) to zero; in this case, the fouling covers the heat transfer surface.

2.3 Effect of Bulk and Surface Temperature

Surface and bulk temperatures have a strong effect on the crystallization fouling resistance; the main reason is an inverse dependence of the supersaturation of these soluble salts with temperature [82]. Various opinions on the effect of temperature on fouling have been reported in literature. This review paper implies "positive, negative or no effect of temperature on the fouling resistance". These different effects aroused great interest among researchers to study the effect of temperature. In a fluidized bed crystallizer, the temperature effect on crystallization investigated by Budz et al. [83]. They showed that increasing the temperature reduces the rate of nucleation. Unlike them, Chenoweth [84] stated that fouling is reduced at low temperatures, and these depositions are usually easily removed. Amjad [85] also confirmed this result; he showed that increasing the temperature increases the reaction rate, more crystal formation, more corrosion, and more fouling.

In subcooled flow boiling conditions, bubbles play a determinant role. At low wall temperatures, due to the low bubble generation frequency, the bubble's contribution to fouling formation is small, and increasing Reynolds number (Re) increases the mass transfer of crystals into the boundary layer, and fouling process is controlled by mass transfer. At high wall temperatures, the bubble generation frequency is high, the bubbles repel the salt crystals inside boundary layer, and fouling process is controlled by reaction. In addition, in mass transfer-controlled region, there is a linear relationship between the fouling resistance and the surface temperature, but in reaction-controlled region, there is an exponential relationship between them [37, 44, 61, 86].

In order to investigate of temperature effect (bulk and surface) on induction period, data from various researches at the early times of experiments have been extracted. Almost all the reviewed studies have the same view on the effect of temperature on induction period length. They showed that an increase in the surface temperature leads to an increase in supersaturation near the heat transfer surface and also increases the reaction rate constant, as a result, the length of the induction period decreases with increasing the surface temperature [19, 61].

Bansal and Müller-Steinhagen [81] investigated the calcium sulfate crystallization fouling in a plate heat exchanger (PHE). Figure 2.7 shows the bulk temperature (T_b) effect on induction period of calcium sulfate crystallization fouling. As shown in Fig. 2.7, at $T_b = 85$ °C, the induction period was about 75 min, after this time the fouling resistance (R_f) became positive, although it did not yet increase rapidly. The induction period is much shorter and reached about 35–40 min for $T_b = 90$ °C, after this time, fouling resistance increased sharply. This indicates that induction period of $CaSO_4$ salt is somewhat dependent on the fluid bulk temperature and with increasing the bulk temperature, the induction period decreases. Mullin [87] also confirmed these results where he described the relationship between the induction time and the bulk and surface temperatures. Kazi et al. [3] and Alahmad [70] showed that induction period length is effectively reduced by bulk temperature increasing as well as increasing the surface supersaturation, due to the accelerated formation of $CaSO_4$ nuclei required to start deposition in the fouling process. Contrary to these studies, Pääkkönen et al. [61], by examining $CaCO_3$ salt, stated that crystallization fouling resistance is slightly affected by bulk temperature; however, although more particles are available in solution at higher bulk temperatures and it increases the probability of particle deposition, weakens the deposition, and increases the removal rate.

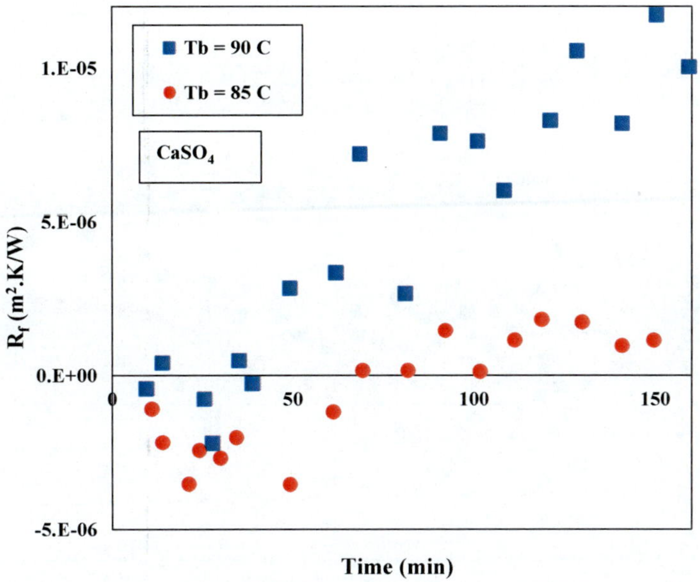

Fig. 2.7 Bulk temperature effect on induction period of $CaSO_4$ fouling (adapted from [81])

Najibi et al. [22] investigated the formation of calcium carbonate deposition during subcooled boiling. They studied surface temperature effect changes by keeping the velocity (0.9 m/s), bulk temperature (T_b = 80 °C) and ion concentration ($[Ca^{2+}]$ = 0.0065 M) constant. The results of this study for the induction period are presented in Fig. 2.8. As can be seen, the induction period can be seen for T_s = 106 °C and to some extent for T_s = 111 °C, but at T_s = 113 °C there is no induction period at all, and the positive fouling resistance is observed from the beginning of the experiment. As the surface temperature increases, the induction period decreases rapidly. They showed that at this flow velocity (0.9 m/s), the deposition rate increases linearly with temperature; hence, the process of fouling is controlled by mass transfer. In addition, they stated that bulk temperature has only a small effect on the induction period and has little effect on the fouling growth period. Abd-Elhady and Malayeri [88] stated that as heated tube surface temperature exceeded the solution saturation temperature by more than 5 °C, e.g., T_s = 110 °C, then the steam bubbles started to appear, and the presence of this bubble and change in the heat transfer mode to subcooled flow boiling increases the fouling and then reduces the induction period length.

Hasson et al. [80] previously presented similar results. Another reason for this phenomenon can be found in the studies of Alahmad [70] and Han et al. [74] where they stated that due to the inverse solubility of $CaSO_4$ with temperature, its solubility decreases with increasing temperature, two main factors that have led to the acceleration of nucleation and fouling production as well as reduction the induction period

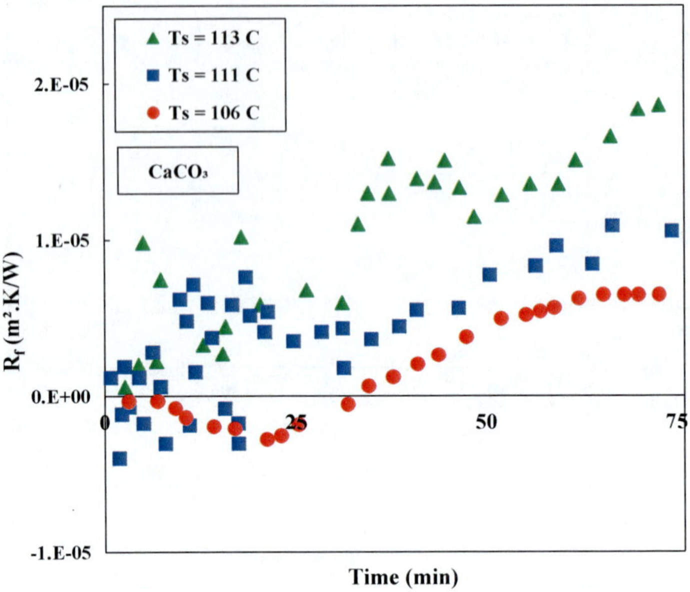

Fig. 2.8 Temperature effect on induction period of $CaCO_3$ fouling (adapted from [22])

length are the increase in the supersaturation level and flow bulk temperature. On the other hand, surface temperature has a direct relationship with the fouling rate, and in fact, the fluid velocity firstly affects the surface temperature.

As the flow velocity increases, due to the increase in flow turbulence, the rate of heat transfer from the surface increases and as a result, the surface temperature decreases. Then, this temperature affects the fouling reaction rate constant according to Eq. (2.1):

$$K_r = k_{r0} \exp\left(-\frac{\Delta E}{RT}\right) \tag{2.1}$$

where k_{r0} is the pre-exponential constant, ΔE is the activation energy, R is the universal gas constant, and T is the absolute temperature at the solid–liquid interface [62].

In addition, longer induction period of scale formation at 106 °C in comparison to that at 113 °C was reported in Fig. 2.9. It indicates that the induction period of fouling depends strongly on the temperature. Similar results were reported by Mullin [87] where he expressed the relationship between the induction period and temperature as Eq. (2.2):

$$\log(T_{int} - 1) = A_1 - \frac{E_a}{2.303\,RT} \tag{2.2}$$

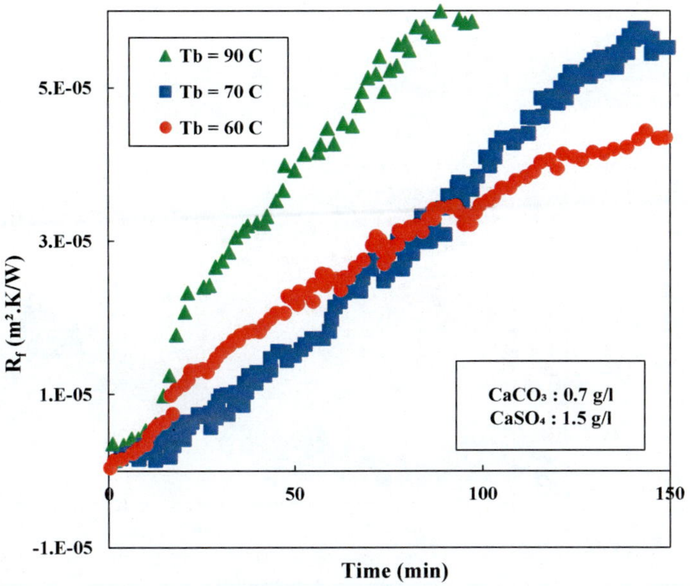

Fig. 2.9 Bulk temperature effect on salt mixture induction period (adapted from [79])

where T_{int} is induction time, E_a is the molar activation energy for nucleation (J/mol), R is the universal gas constant, and A_1 is a constant value.

Helalizadeh et al. [79] studied salt mixtures crystallization fouling under subcooled flow boiling and convective heat transfer. Figures 2.9 and 2.10 show the effect of bulk temperature (T_b) and heated surface temperature (T_s), respectively at constant flow velocity and concentrations of $[CaSO_4] = 1.5$ g/l and $[CaCO_3] = 0.7$ g/l. The operating conditions used in this research are $[CaCO_4]$: 1–2.5 g/l, $[CaCO_3]$: 0.25–1 g/l, u: 0.5–2 m/s and $\dot{q}$: 100–400 kW/m^2. As shown in Fig. 2.9, no induction period was observed at three temperatures studied. The deposition was formed quickly, and the fouling resistance was positive from the beginning of the process, and increased linearly at all temperatures.

In Fig. 2.10, where the data obtained at a constant flow velocity (0.5 m/s) and bulk temperature ($T_b = 80$ °C) to investigate the surface temperature changes ($T_s = 103$–113 °C), no induction period was observed. However, a remarkable point is the flow regimes effect on fouling resistance curves, which is very different at $T_s = 103$ °C (forced convective heat transfer) with the curves obtained at $T_s = 107$ and 113 °C (subcooled flow boiling) due to the presence of bubbles.

In most of the reviewed studies, it was clearly stated that increasing the surface temperature decreases the induction period length. This result can also be seen in other studies [19, 21, 37, 43, 65, 69, 78, 80, 87]. However, increasing the bulk temperature has less effect on the induction period of crystallization fouling [19, 22, 29, 37, 43, 65, 68, 74, 78, 80, 87].

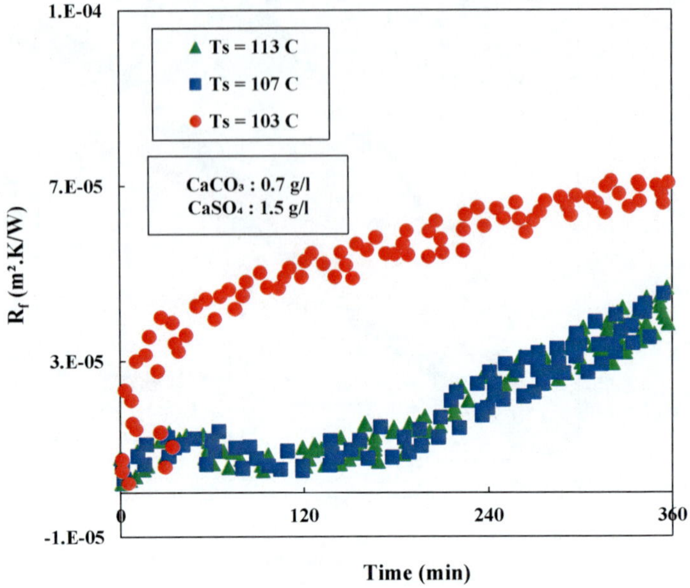

Fig. 2.10 Surface temperature effect on salt mixture induction period (adapted from [79])

In general, regarding the effect of temperature on the induction period, it can be concluded that due to the inverse solubility of $CaSO_4$ and $CaCO_3$ salts with temperature, the crystallization fouling of these salts is strongly affected by temperature. Therefore, the length of the induction period decreases significantly with increasing the surface temperature due to increasing surface supersaturation, and acceleration of nucleation in the fouling process. However, bulk temperature effect is not the same as the surface temperature. The temperature of the bulk fluid has no effect on the induction period both at low velocity (where mass transfer is important) and at high velocity (where surface reaction is important).

2.4 Effect of Salt Concentration

Studies on salt concentration effect on induction period are very scarce. Therefore, in order to investigate the effect of this parameter on the induction period, among the existing studies, data related to the initial times of experiments were extracted and reviewed separately. Among the various studies, the data of the three studies were presented in the early experimental times for $CaSO_4$ salt under forced convection conditions, for $CaCO_3$ salt, and mixed salt ($CaSO_4$ and $CaCO_3$) under subcooled flow boiling conditions, respectively.

Müller-Steinhagen [65] investigated the effect of salt concentration on $CaSO_4$ fouling using three different concentrations in a plate heat exchanger (PHE). He provided the fouling data up to about 6000 min, but here, only data at different concentrations up to the first 600 min of the experiment were considered to investigate the induction period. A 1 μm filter was placed in the flow path to prevent the particles fouling. The results are presented in Fig. 2.11. It can be seen that at the concentration C = 2.8263 g/l, the amount of negative fouling resistance continued even up to 600 min after the start of the experiment, but with increasing the concentration to C = 2.9798 g/l, after about 250 min, the negative fouling resistance ended. For a concentration of C = 3.1819 g/l this occurred at about 200 min. Therefore, with increasing the bulk concentration due to higher supersaturation, the induction period became shorter.

In another study, Mwaba et al. [16] stated that under forced convective heat transfer conditions, the induction period and degree of supersaturation had a relatively strong dependence. At $C_{b,\,init} = 1.4\ C_s$, the initial period was relatively long (4000 min) while at $C_{b,\,init} = 1.7\ C_s$, the induction period was almost non-observable. At high supersaturation values, the nucleation rate increases and thus the length of the induction period decreases. In addition, high supersaturation leads to the bulk crystallization phenomenon, which finally produces particulate fouling characterized by the absence of an induction period. They also showed that in about 40% supersaturation there is no long induction period and spontaneous crystallization, while in 60–70% supersaturation the results were completely opposite. The results of Alahmad [70] also confirmed the results of the two previous studies. He stated that as the $CaSO_4$

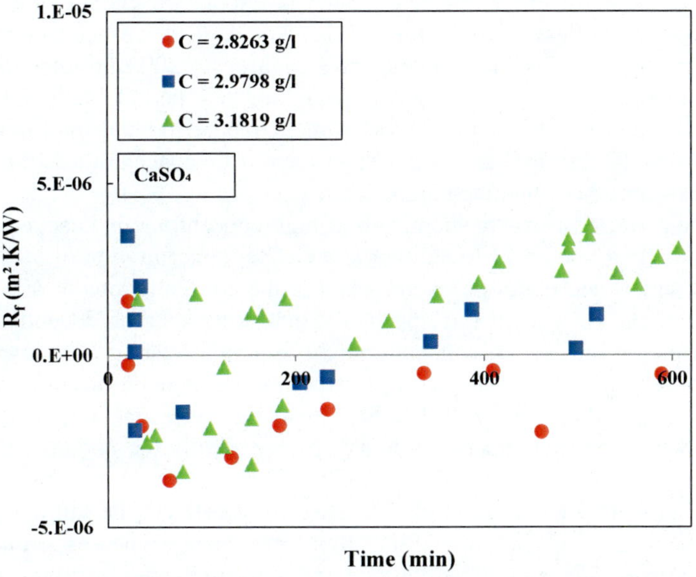

Fig. 2.11 Concentration effect on induction period of $CaSO_4$ fouling (adapted from [65])

concentration in fluid increased, the deposition increased sharply. Since the driving force behind fouling formation is salt supersaturation, an increase in concentration will definitely lead to an increase in fouling formation. In addition, he showed in his experiments that by tripling the salt concentration (500–1500 ppm), the induction period decreased from 168 to 82 h. This reduction can also be attributed to $CaSO_4$ supersaturation that leading to the early scale nuclei creation and initiating the scale process on the surface. It should be noted that this study was also performed under forced convective conditions. Jalalirad et al. [33] have investigated the possibility of using projectiles in tubular heat exchangers of water service systems as a fouling mitigation method. They showed that by increasing concentration from 3 to 5 g/l, the induction period length decreased from 8 h to no induction time due to supersaturation. The fouling rate is usually expressed in the following form:

$$\dot{n} = k\left(C_b - C^*\right)^n \tag{2.3}$$

where $\dot{n}$ is fouling rate, k is constant reaction rate, C_b is bulk concentration and C^* is interface concentration. The values of 1 and 2 for n are assigned in the controlled fouling process by mass transfer and chemical reaction, respectively [62]. Therefore, concentration should have an important effect on the fouling process.

In order to study the concentration effect on $CaCO_3$ induction period, the data presented in the study of Najibi et al. [22] were used at the first 200 min of the experiment (total test time was 600 min). Figure 2.12 shows the results of this study. This study was performed under subcooled flow boiling conditions. Surface temperature, bulk temperature and flow velocity were assumed to be constant at $T_s = 112$ °C, $T_b = 80$ °C and u = 1 m/s, respectively. They stated that the effect of concentration is very strong regardless of the fouling mechanism. At the lowest concentration, i.e., 3.7 g/l, the fouling resistance was negative even up to 200 min after the start of the experiment and there was a relatively long induction period, but with increasing the concentration to 6.9 g/l, there was no induction period and fouling resistance increased rapidly after the start of the experiment. At a concentration of 8.2 g/l, the fouling resistance became much higher.

The main cause of this phenomenon at high concentrations is supersaturation. In addition, Teng et al. [68] found that at a $CaCO_3$ concentration of 500 mg/l, the induction period was about one hour, while at the concentrations of 400 mg/l and 300 mg/l, it was 7 h and 14 h, respectively, indicating a strong dependence of the induction period on the salt concentration. The similar results were also reported by Xiaokai et al. [69]. In general, supersaturation is the main driving force in the crystallization fouling process. Therefore, with increasing the salt concentration, initial nucleation of salts, supersaturation and consequently, the amount of deposition increase and the induction time decreases [37, 63, 87].

The results of Helalizadeh et al. [79] used to investigate the salt concentration effect on $CaSO_4$ and $CaCO_3$ mixed salts fouling resistance. Their experiments were performed under subcooled flow boiling conditions. In the present study, data were used at the initial 100 min of the experiment (the total length of the experiment was 1000 min). Flow velocity, heat flux and bulk temperature are assumed to be constant

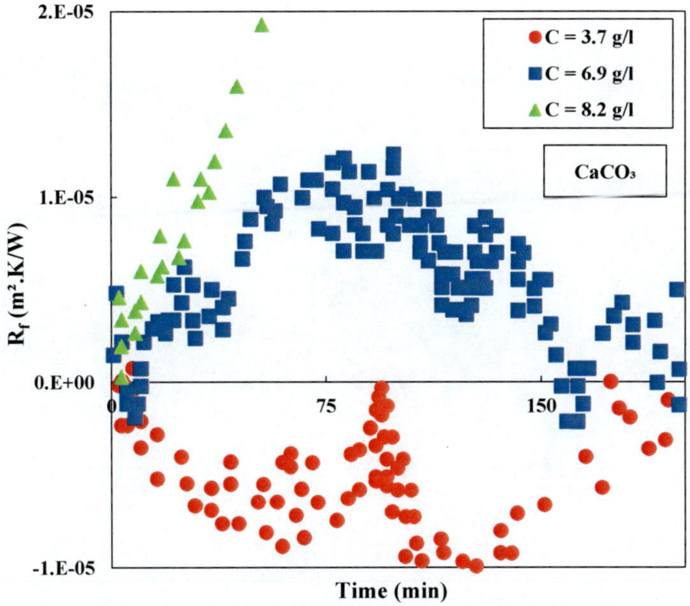

Fig. 2.12 Concentration effect on $CaCO_3$ fouling induction period (adapted from [22])

at u = 0.5 m/s, $\dot{q}$ = 300 kW/m^2 and T_b = 80 °C, respectively. Figure 2.13 shows three different concentrations from this experiment.

It can be clearly seen that at all three concentrations, no induction period was observed. Bulk composition had a strong effect on the fouling rates. The important factor relevant to crystallization fouling is the degree of supersaturation of the deposit forming species, rather than the molar concentration. Sudmalis and Sheikholeslami [76] showed that with increasing the initial calcium concentration, the induction period for pure $CaCO_3$ reduced. For pure $CaSO_4$ and mixed solutions, the effect of concentration increase on the induction time was not noticeable. In mixed salt experiments, the precipitate structure was affected by co-existence of the salts, and the fouling layer was stronger than pure $CaSO_4$ and weaker than pure $CaCO_3$ precipitate.

Sheikholeslami [75] proved that the induction period in the mixed solutions followed the same time pattern as that observed in pure $CaCO_3$ solutions. It suggests that, once the $CaCO_3$ is formed, it then acts as nuclei for the formation of $CaSO_4$. She also showed that induction period is salt specific. The induction period length for $CaSO_4$ is much longer than that of $CaCO_3$ in single salt crystallization. It also depends on the degree of supersaturation of the precipitating salt. Their results showed that the presence of other soluble and crystallizing species does indeed affect the induction period.

As clearly shown in the reviewed studies, for pure salts, regardless of the type of salt, the induction period decreases with increasing the salt concentration under both heat transfer conditions [3, 22, 43, 68, 74, 78, 80, 81, 88, 89]. However, in the case

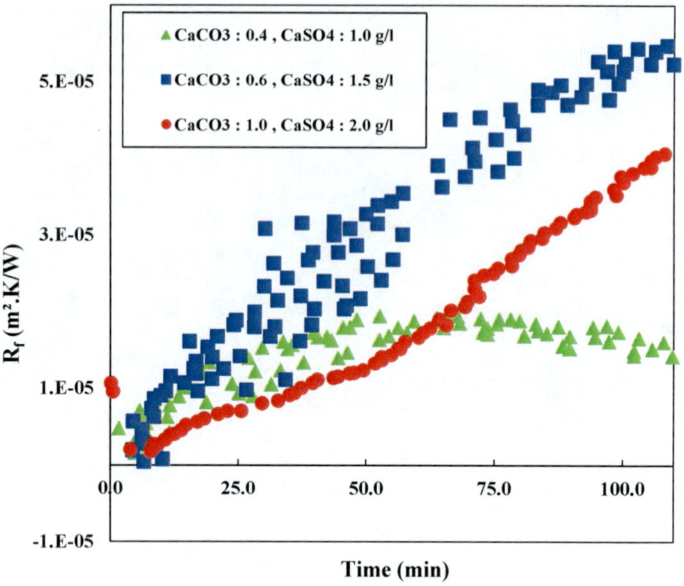

Fig. 2.13 Concentration effect on induction period of mixed salts fouling (adapted from [79])

of salt mixtures, as shown in Fig. 2.13, there is no induction period [79]. However, due to the lack of sufficient studies on the mixed salt, providing an accurate result requires further investigation.

2.5 Effect of Heat Flux

It should be mentioned that no unique research has been found on the effect of heat flux on the induction period. To investigate this parameter, according to the previous sections on the induction period, experimental data at the beginning of the experiment from different studies have been used. Results of Esawy [90] and Esawy et al. [91] are used under pool boiling conditions for $CaSO_4$ salt, from first 60 and 500 min of the experiment, respectively. In addition, data from Pääkkönen et al. [61] also used for $CaCO_3$ salt under forced convective conditions at the first 100 min of the experiment.

Esawy [90] investigated the effect of heat flux on the fouling resistance at a constant concentration of $CaSO_4$. This test was performed at three different fluxes (100, 200 and 300 kW/m^2) on the surface of a stainless-steel tube. The operating conditions of the study are C_b: 1.2–1.6 g/l, T_s: max 170 °C, T_b: 101 °C, q: 5–300 kW/m^2. Figure 2.14 shows the results of this study. The fouling resistance was

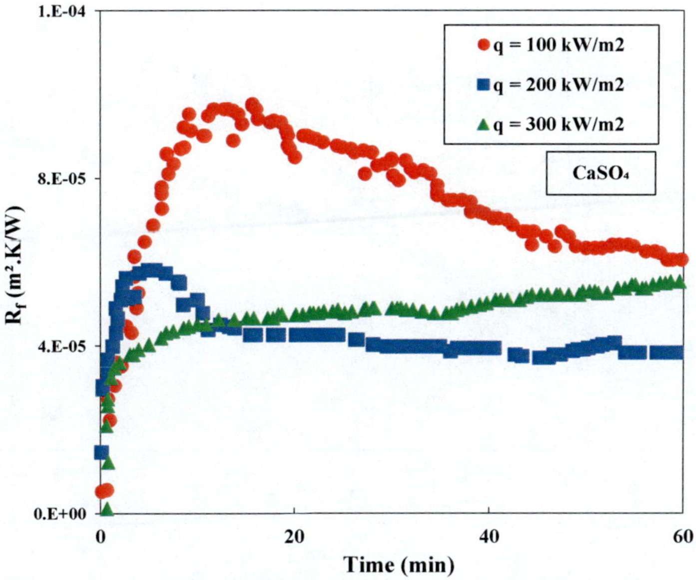

Fig. 2.14 Heat flux effect on induction period of $CaSO_4$ salt fouling (adapted from [90])

varied strongly with time as heat flux was augmented. At these three heat fluxes, no induction period was observed, and the fouling resistances at all three heat fluxes were in positive region from the beginning of the test and did not become negative at all.

In another study on pool boiling heat transfer conditions, Esawy et al. [91] investigated the effect of sintering on $CaSO_4$ fouling formation. They measured the fouling resistance at three different fluxes in their research. Figure 2.15 shows the effect of heat flux on the induction period at these three heat fluxes. As shown, no significant induction period was observed in all the experiments. Sarafraz et al. [49–56] by examining the heat transfer coefficient of different fluids under pool boiling stated that the frequency of bubble formation depends on the heat flux. They showed that the deposition of particles on the surface significantly reduces the heat transfer coefficient and therefore, creates resistance on the heat transfer surface. In addition, increasing the heat flux greatly increases this fouling resistance.

Furthermore, Malayeri et al. [89] showed that the length of the induction period in $CaSO_4$ fouling decreased with increasing heat flux. This result was attributed to the increase in the surface temperature with increasing heat flux. Primary nucleation plays an important role in the fouling process, as long as the removal process is negligible; the initial nucleation accelerates the formation of scale on the heat transfer surface. Increasing the heat flux accelerates this phenomenon and causes a linear increase in the fouling resistance. They showed that by increasing the heat flux from 33 to 300 kW/m^2, the induction time decreased from about 1300 min to only 20 min. Later, Malayeri and Müller-Steinhagen [32] confirmed the results of this study. In

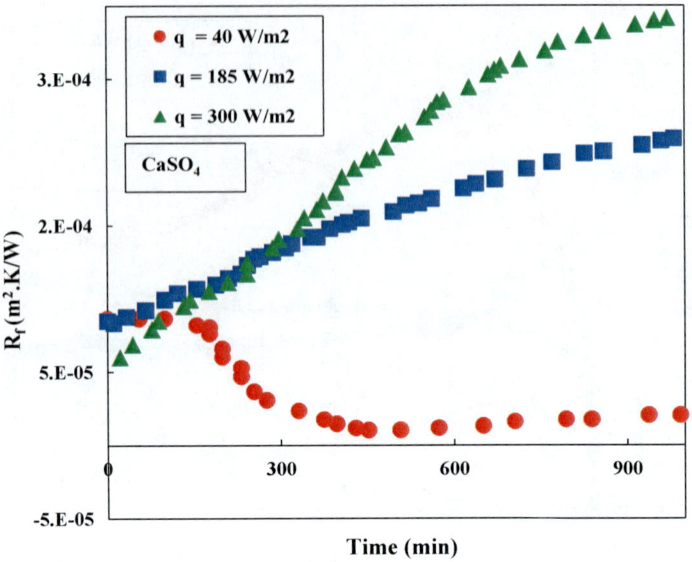

Fig. 2.15 Heat flux effect on induction period of $CaSO_4$ salt fouling (adapted from [91])

another study on $CaSO_4$ deposition, Herz et al. [24] have stated that this result can be attributed to the increase in surface temperature as heat flux increases; hence, the solubility limit of calcium sulphate ions is exceeded further, promoting early nucleation. The earlier nucleation plays an important role and leads to faster fouling of the heat transfer surface as long as the removal rate is negligible. Evidently, the fouling rate depends almost linearly on the heat flux.

Pääkkönen et al. [61] investigated the crystallization fouling of $CaCO_3$ under forced convective conditions. The effect of heat flux on the fouling resistance during the induction period is shown in Fig. 2.16. They also showed that the induction period decreased with increasing heat flux. As the heat flux increased, the wall temperature increased, which increased the crystallization rate due to the effect of temperature on the reaction rate constant and supersaturation. Therefore, a short induction period was seen at a higher heat flux. Yang et al. [19] reported similar results previously. In addition, Fig. 2.16 showed that by increasing the heat flux from 53 to 62 kW/m^2, the induction period decreased from about 100 min to less than 30 min. This indicates the inverse relationship between the induction time and heat flux, a phenomenon that has been observed in other studies [23, 44, 72, 86, 87, 92].

In general, increasing the heat flux, regardless of the type of salt, reduced the induction period length. The cause of this phenomenon is an increase in wall temperature, which causes an increase in supersaturation due to the inverse solubility of the salts. Another reason for this phenomenon is the increase in nucleation and the number of active bubble sites with increasing the heat flux; fouling is formed beneath these bubbles.

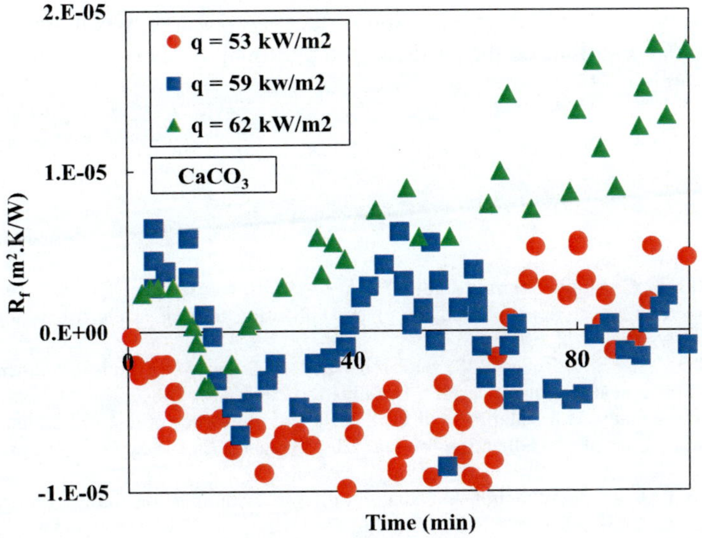

Fig. 2.16 Heat flux effect on induction period of $CaCO_3$ salt fouling (adapted from [61])

2.6 Concluding Remarks

In the present chapter, a comprehensive review of the effect of various operational parameters on the induction period of crystallization fouling of calcium salts over the past few decades is presented. The following results are obtained:

- Very rare information is available on the effect of operating parameters on the induction period. Extensive studies are needed in this area.
- The effect of pH on the induction period as well as the effect of operating parameters on the induction period of calcium mixed salt has rarely been found in reviewed studies.
- The effect of flow velocity is complex. When the effects of additional disturbances that are directly related to velocity changes are greater than the effects of heat transfer area reduction, the increase in heat transfer coefficient overrides its reduction due to fouling, and fouling resistance becomes negative and the induction period increases.
- With increasing the surface temperature due to increasing surface supersaturation, and acceleration of nucleation in the fouling process, the length of the induction period decreases significantly. However, bulk temperature has a negligible effect on the induction period.
- The induction period decreases with increasing the salt concentration. However, in the case of salt mixtures, there is no induction period.

- Increasing the heat flux increases the surface temperature, which leads to supersaturation conditions on the surface, which ultimately leads to a reduction in the induction period.

References

1. R.B. Ritter, "Crystalline fouling studies," Trans. ASME, vol. 105, pp. 374–378, 1983.
2. A. Janzen and E.Y. Kenig, "Analysis of crystallization fouling in electric water heating," Heliyon, vol. 5, no. 11, e02695.
3. Kazi SN, Teng KH, Zakari MS, Sadeghinezhad E, Bakar MA. Study of mineral fouling mitigation on heat exchanger surface. Desalination 2015; 367: 248–254.
4. M.S. Abd-Elhady, M.R. Jalalirad and M.R. Malayeri, "Influence of injected projectiles on the induction period of crystallization fouling," Heat Transfer Eng., vol. 35, no. 3, pp. 232–245, 2014.
5. Geddert T, Augustin W, Scholl S. Induction time in crystallization fouling on heat transfer surfaces. Chem Eng Technol 2011; 34(8): 1303–1310.
6. Bansal B, Müller-Steinhagen H, Chen XD. Effect of suspended particles on crystallization fouling in plate heat exchangers. J Heat Trans 1997; 119(3): 568–574.
7. F. Wang, T.E. Davis and V.V. Tarabara, "Crystallization of Calcium Sulfate Dihydrate in the Presence of Colloidal Silica," Ind. Eng. Chem. Res. vol. 49, pp. 11344–11350, 2010.
8. Y. Chuanfang, "Mechanisms of Calcium Carbonate Scaling During and Post Induction Period," in Proc. Int. Conf. on Fouling Mitigation of Industrial Heat-Exchange Equipment, San Luis Obispo, CA, June 1995, pp. 275–286.
9. W.T. Kim, C. Bai and Y.I. Cho, "A study of $CaCO_3$ fouling with a microscopic imaging technique," Int. J. Heat Mass Tran., vol. 45, no. 3, pp. 597–607, 2002.
10. Albert, F., Augustin, W., Scholl, S., 2009, Enhancement of Heat Transfer in Crystallization Fouling due to Surface Roughness, in Proceedings of Eurotherm Conference on Fouling and Cleaning in Heat Exchangers, Schladming, Austria, June 14–19, pp. 303–310.
11. M.S. Abd-Elhady and M.R. Malayeri, "Impact of hardness and surface texture on cleaning action of various projectiles," Chem. Eng. Res. Des., vol. 94, pp. 153–163, 2015.
12. M. Yang, A. Young, A. Niyetkaliyev and B. Crittenden, "Modelling the fouling induction period," Int. J. Therm. Sci., vol. 51, pp. 175–183, 2012.
13. Peyghambarzadeh SM, Bahrami N. Statistical analysis of calcium sulfate scaling under boiling heat transfer. Appl Therm Eng 2013; 53: 108–113.
14. Geddert T, Bialuch I, Augustin W, Scholl S. Extending the induction period of crystallization fouling through surface coating. Heat Transfer Eng 2009; 30(10–11): 868–875.
15. F. Fahiminia, A.P. Watkinson and N. Epstein, "Early events in the precipitation fouling of calcium sulphate dihydrate under sensible heating conditions," Can. J. Chem. Eng., vol. 85, no. 5, pp. 679–691, 2007.
16. Mwaba MG, Rindt CCM, Van Steenhoven AA, Vorstman MAG. Experimental investigation of $CaSO_4$ crystallization on a flat plate. Heat Transfer Eng 2006; 27(3): 42–54.
17. G. Rizzo, H. Müller-Steinhagen and E. Richter, "Induction period of heterogeneous nucleation during crystallization fouling: ion implantation effects," in Proc. Int. Conf. on Heat Exchanger Fouling and Cleaning VI., Kloster Irsee, Germany, June 5–10 2005, pp. 316–326.
18. Fahiminia, F., Watkinson, A. P., Epstein, N., 2007, Calcium Sulfate Scaling Delay Times Under Sensible Heating Conditions, in Proceedings of the 6th International Conference on Fouling and Cleaning in Heat Exchangers, Tomar, Portugal, July 1–6, ECI Symposium Series, RP2, pp. 176–184.
19. Q. Yang, Y. Liu, A. Gu, J. Ding and Z. Shen, "Investigation of induction period and morphology of $CaCO_3$ fouling on heated surface," Chem. Eng. Sci., vol. 57, no. 6, pp. 921–931, 2002.

20. Z.S. Saleh, R. Sheikholeslami and A.P. Watkinson, "Fouling characteristics of a light Australia crude oil," Heat Tran. Eng., vol. 26, no. 1, pp. 15–22, 2005.
21. Q. Zhenhua, C. Yongchang and M. Chongfang, "Experimental study of fouling on heat transfer surface during forced convective heat transfer," Chinese J. Chem. Eng., vol. 16, no. 4, pp. 535–540, 2008.
22. Najibi, S.H., Müller-Steinhagen, H., Jamialahmadi, M., 1997, Calcium Carbonate Scale Formation during Subcooled Flow Boiling, Journal of Heat Transfer, 119, 767–775.
23. Q.F. Yang, D.Q. Xu, and Z.Q. Shen, "A Theoretical Analysis and Experimental Study of the Induction Period of Calcium Carbonate Scaling," J. Chem. Ind. Eng. (China), vol. 45, pp. 199–205, 1994.
24. A. Herz, M.R. Malayeri and H. Müller-Steinhagen, "Fouling of roughened stainless steel surfaces during convective heat transfer to aqueous solutions," Energy Conv. Manag.; vol. 49, no. 11, pp. 3381–3386, 2008.
25. A. Stärk, K. Krömer, K. Loisel, K. Odiot, S. Nied and H. Glade, "Impact of tube surface Properties on Crystallization Fouling in Falling Film Evaporators for Seawater Desalination," Heat Tran. Eng., vol. 38, no. 7–8, pp. 762–774, 2016.
26. M.R. Malayeri, A. Al-Janabi, and H. Müller-Steinhagen, "Application of nano-modified surfaces for fouling mitigation," Int. J. Energy Res., vol. 33, no. 13, pp. 1101–1113, 2009.
27. M. Förster, and M. Bohnet, "Modification of molecular interactions at the interface crystal/heat transfer surface to minimize heat exchanger fouling," Int. J. Therm. Sci., vol. 39, pp. 697–708, 2000.
28. F. Albert, W. Augustin and S. Scholl, "Roughness and constriction effects on heat transfer in crystallization fouling," Chem. Eng. Sci., vol. 66, no. 3, pp. 499–509, 2011.
29. Förster M, Augustin W, Bohnet M, Influence of the adhesion force crystal/heat exchanger surface on fouling mitigation. Chem Eng Process 1999; 38: 449–461.
30. Müller-Steinhagen, H., Malayeri, M.R., Watkinson, A.P., 2011, Heat Exchanger Fouling: Mitigation and Cleaning Strategies, Heat Transfer Engineering, 32(3–4), 189–196.
31. M. Mayer, W. Augustin, and S. Scholl, "An approach to modeling induction period in crystallization fouling," Heat Mass Transfer, vol. 49, pp. 1419–1432, 2013.
32. Malayeri, M. R., Müller-Steinhagen, H., 2007, Initiation of $CaSO_4$ Scale Formation on Heat Transfer Surface under Pool Boiling Conditions, Heat Transfer Engineering, 28(3), 240–247.
33. M.R. Jalalirad, M.R. Malayeri and R. Preimesser, "Online cleaning of tubular heat exchangers in water service systems using projectiles," Desalination and Water Treatment, vol. 51, no. 4–6, pp. 780–785, 2012.
34. A. Al-Janabi, M.R. Malayeri and H. Müller-Steinhagen, "Experimental investigation of crystallization fouling on grooved stainless steel surfaces during convective heat transfer," Heat Transfer Eng., vol. 30, no. 10–11, pp. 832–839, 2009.
35. H. Föste, F. Stehmann, W. Augustin and S. Scholl, "Pulsed flow and surface coatings to mitigate fouling," in Proc. Eurotherm Int. Conf. on heat exchanger fouling and cleaning X., Crete, Greece, 9–14 June 2013, pp. 397–402.
36. A. Al-Janabi, M.R. Malayeri and O.O. Badran, "Performance of shot peened surfaces subject to crystallization fouling," Int. J. Therm. Sci., vol. 111, pp. 379–389, 2017.
37. Peyghambarzadeh, S.M., Vatani, A., Jamialahmadi, M., 2013, Influences of bubble formation on different types of heat exchanger fouling, Applied Thermal Engineering, 50, 848–856.
38. M.M. Sarafraz, S.M. Peyghambarzadeh and N. Vaeli, "Subcooled flow boiling heat transfer of ethanol aqueous solution in vertical annulus space," Chem. Ind. Chem. Eng. Q., vol. 18, no. 2, pp. 315–327, 2012.
39. S.M. Peyghambarzadeh, M.M. Sarafraz, N. Vaeli, E. Ameri, A. Vatani and M. Jamialahmadi, "Forced convective and subcooled flow boiling heat transfer to pure water and n-heptane in an annular heat exchanger," Ann. Nucl. Energy, vol. 53, pp. 401–410, 2013.
40. M.M. Sarafraz and S.M. Peyghambarzadeh, "Experimental study on subcooled flow boiling heat transfer to water–diethylene glycol mixtures as a coolant inside a vertical annulus," Exp. Therm. Fluid Sci., vol. 50, pp. 154–162, 2013.

41. A. Zangeneh, A. Vatani, Z. Fakhroeian and S.M. Peyghambarzadeh, "Experimental study of forced convection and subcooled flow boiling heat transfer in a vertical annulus using different novel functionalized ZnO nanoparticles," Appl. Therm. Eng., vol. 109, Part A, pp. 789–802, 2016.
42. A. Zangeneh, S.M. Peyghambarzadeh, A. Bayat and A. Vatani, "Application of general multi-level factorial design approach in forced convection and subcooled flow boiling heat transfer to CuO/water nanofluids," J. Mol. Liq., vol. 313, 113502, 2020.
43. S.M. Peyghambarzadeh, A. Vatani and M. Jamialahmadi, "Application of asymptotic model for the prediction of fouling rate of calcium sulfate under subcooled flow boiling," Appl. Therm. Eng., vol. 39, pp. 105–113, 2012.
44. A. Vosough, S.M. Peyghambarzadeh and M.R. Assari, "Influence of thermal shock on the mitigation of calcium sulfate crystallization fouling under subcooled flow boiling condition," Appl. Therm. Eng., vol. 164, 114434, 2020.
45. A. Shaltouki and S.M. Peyghambarzadeh, "Forced convective and subcooled flow boiling heat transfer to water-Al_2O_3 microfluid in an annular heat exchanger," IJST-T Mech. Eng., vol. 43, no. 25–26, pp. 869–879, 2019.
46. V. Nikkhah, M.M. Sarafraz, F. Hormozi and S.M. Peyghambarzadeh, "Particulate fouling of CuO-water nanofluid at isothermal diffusive condition inside the conventional heat exchanger-experimental and modeling," Exp. Therm. Fluid Sci., vol. 60, pp. 83–95, 2014.
47. S.M. Peyghambarzadeh, A. Vatani and M. Jamialahmadi, "Experimental study of micro-particle fouling under forced convective heat transfer," Braz. J. Chem. Eng., vol. 29, no. 4, pp. 713–724, 2012.
48. M.M. Sarafraz, F. Hormozi and S.M. Peyghambarzadeh, "Role of nanofluid fouling on thermal performance of a thermosyphon: Are nanofluids reliable working fluid?" Appl. Therm. Eng., vol. 82, pp. 212–224, 2015.
49. M.M. Sarafraz, F. Hormozi and S.M. Peyghambarzadeh, "Pool boiling heat transfer to aqueous alumina nano-fluids on the plain and concentric circular micro-structured (CCM) surfaces," Exp. Therm. Fluid Sci., vol. 72, pp. 125–139, 2015.
50. M.M. Sarafraz, S.M. Peyghambarzadeh and S.A. Alavifazel, "Enhancement of nucleate pool boiling heat transfer to dilute binary mixtures using endothermic chemical reactions around the smoothed horizontal cylinder," Heat Mass Transfer, vol. 48, pp. 1755–1765, 2012.
51. M.M. Sarafraz, S.M. Peyghambarzadeh and S.A. Alavifazel, "Enhancement of the pool boiling heat transfer coefficient using the gas injection into the water," Polish J. Chem. Technol., vol. 14, no. 4, pp. 100–109, 2012.
52. M.M. Sarafraz, S.M. Peyghambarzadeh and S.A. Alavifazel, "Experimental studies on nucleate pool boiling heat transfer to ethanol/MEG/DEG ternary mixture as a new," Chem. Ind. Chem. Eng. Q., vol. 18, no. 4, pp. 577–586, 2012.
53. M.M. Sarafraz and S.M. Peyghambarzadeh, "Influence of thermodynamic models on the prediction of pool boiling heat transfer coefficient of dilute binary mixtures," Int. Commun. Heat Mass, vol. 39, pp. 1303–1310, 2012.
54. M.M. Sarafraz, S.M. Peyghambarzadeh, S.A. Alavifazel and N. Vaeli, "Nucleate pool boiling heat transfer of binary nano mixtures under atmospheric pressure around a smooth horizontal cylinder," Period. Polytech. Chem. Eng., vol. 57, no. 1–2, pp. 71–77, 2013.
55. M.M. Sarafraz and S.M. Peyghambarzadeh, "Nucleate pool boiling heat transfer to Al_2O_3-Water and TiO_2-water nanofluids on horizontal smooth tubes with dissimilar homogeneous materials," Chem. Biochem. Eng. Q., vol. 26, no. 3, pp. 199–206, 2012.
56. M.M. Sarafraz, F. Hormozi, S.M. Peyghambarzadeh and E. Salari, "Experimental study on the influence of SO_2 gas injection to pure liquids on pool boiling heat transfer coefficients," Heat Mass Transfer, vol. 50, pp. 747–757, 2014.
57. S.M. Peyghambarzadeh, A. Hatami, A. Ebrahimi and S.A. Alavifazel, "Photographic study of bubble departure diameter in saturated pool boiling to electrolyte solution," Chem. Ind. Chem. Eng. Q., vol. 20, no. 1, pp. 143–153, 2014.
58. S.M. Peyghambarzadeh, M. Jamialahmadi, S.A. Alavifazel and S. Azizi, "Saturated nucleate boiling to binary and ternary mixtures on horizontal cylinder," Exp. Therm. Fluid Sci., vol. 33, pp. 903–911, 2009.

59. S.A. Alavifazel, M.M. Sarafraz, A. Arabi Shamsabadi and S.M. Peyghambarzadeh, "Pool boiling heat transfer in diluted water/glycerol binary solutions," Heat Transfer Eng., vol. 34, no. 10, pp. 828–837, 2013.
60. L.C. Wang, S.F. Li, L.B. Wang, K. Cui, Q.L. Zhang, H.B. Liu and G. Li, "Relationships between the characteristics of $CaCO_3$ fouling and the flow velocity in smooth tube," Exp. Therm. Fluid Sci., vol. 74, pp. 143–159, 2016.
61. Pääkkönen, T.M., Riihimäki, M., Simonson, C.J., Muurinen, E., Keiski, R.L., 2012, Crystallization fouling of $CaCO_3$ - Analysis of experimental thermal resistance and its uncertainty, International Journal of Heat and Mass Transfer, 55, 6927–6937.
62. Najibi, S.H., 1996, Heat transfer and heat transfer fouling during subcooled boiling for Electrolyte Solutions, Ph.D. thesis, Department of Chemical and Process Engineering, University of Surrey.
63. Hoang TA, Ang M, Rohl AL. Effects of process parameters on gypsum scale formation in pipes. Chem Eng Technol; 34 (6): 1003–1009.
64. Bansal B, Chen XD, Müller-Steinhagen H. Analysis of 'classical' deposition rate law for crystallization fouling. Chem Eng Process 2008; 47: 1201–1210.
65. Müller-Steinhagen, H., 1999, Cooling-Water Fouling in Heat Exchangers, Advances in Heat Transfer, 33, 415–496.
66. A.P. Watkinson and O. Martinez, "Scaling of heat exchanger tubes by calcium carbonate," J. Heat Transfer, vol. 97, no. 4, pp. 504–508, 1975.
67. P. Walker and R. Sheikholeslami, "Assessment of the effect of velocity and residence time in $CaSO_4$ precipitating flow reaction," Chem. Eng. Sci., vol. 58, no. 16, pp. 3807–3816, 2003.
68. K.H. Teng, S.N. Kazi, A. Amiri, A.F. Habali, M.A. Bakar, B.T. Chewa, A. Al-Shamma, A. Shaw, K.H. Solangi and G. Khan, "Calcium carbonate fouling on double-pipe heat exchanger with different heat exchanging surfaces," Powder Technol., vol. 315, pp. 216–226, 2017.
69. X. Xiaokai, M. Chongfang and C. Yongchang, "Mechanism of calcium carbonate scale deposition under subcooled flow boiling conditions," Chinese J. Chem. Eng., vol. 13, no. 4, pp. 464–470, 2005.
70. M. Alahmad, "Factors affecting scale formation in sea water environments - an experimental approach," Chem. Eng. Technol., vol. 31, no. 1, pp. 149–156, 2008.
71. L.D. Tijing, H.Y. Kim, D.H. Lee, C.S. Kim and Y.I. Cho, "Physical water treatment using RF electric fields for the mitigation of $CaCO_3$ fouling in cooling water," Int. J. Heat Mass Tran. vol. 53, pp. 1426–1437, 2010.
72. L.M. Al-Hadhrami, A. Ahmad and A. Al-Qahtani, "Experimental study of fouling resistance in twisted tube heat exchanger," Heat Transfer Eng., vol. 33, no. 12, pp. 1024–1032, 2012.
73. Hasson, D., Zahavi, J., 1970, Mechanism of Calcium Sulfate Scale Deposition on Heat Transfer Surfaces, I&EC Fundamentals, 9(1), pp. 1–10.
74. Z. Han, Z. Xu and J. Wang, "$CaSO_4$ fouling characteristics on the rectangular channel with half-cylinder vortex generators," Appl. Therm. Eng., vol. 128, pp. 1456–1463, 2018.
75. R. Sheikholeslami, "Nucleation and kinetics of mixed salts in scaling," AIChE J., vol. 49, no. 1, pp. 194–202, 2003.
76. M. Sudmalis and R. Sheikholeslami, "Coprecipitation of $CaCO_3$ and $CaSO_4$," Can. J. Chem. Eng., vol. 78, no. 1, pp. 21–31, 2009.
77. T.H. Chong and R. Sheikholeslami, "Thermodynamics and kinetics for mixed calcium carbonate and calcium sulfate precipitation," Chem. Eng. Sci., vol. 56, no. 18, pp. 5391–5400.
78. Najibi, S.H., Müller-Steinhagen, H., Jamialahmadi, M., 1997, Calcium sulphate scale formation during subcooled flow boiling, Chemical Engineering Science, 52(8), 1265–1284.
79. Helalizadeh, A., Müller-Steinhagen, H., Jamialahmadi, M., 2000, Mixed salt crystallization fouling, Chem. Eng. Process, 39, 29–43.
80. D. Hasson, M. Avriel, W. Resnick, T. Rozenman and S. Winderich, "Mechanism of calcium carbonate scale deposition on heat transfer surfaces," Ind. Eng. Chem. Fund., vol. 7, pp. 59–65, 1968.
81. Bansal, B., Müller-Steinhagen, H., 1993, Crystallization Fouling in Plate Heat Exchangers, ASME Journal of Heat Transfer, 115, pp. 584–591.

82. Zhao X, Chen XD. A critical review of basic crystallography to salt crystallization fouling in heat exchangers. Heat Transfer Eng 2013; 34(8–9): 719–732.
83. J. Budz, P. Karpiński, Z. Nuruć, "Effect of temperature on crystallization and dissolution processes in a fluidized bed," AIChE J., vol. 31, no. 2, pp. 259–268, 1985.
84. J.M. Chenoweth, "Final report of the HTRI/TEMA joint committee to review the fouling section of the TEMA standards," Heat Transfer Eng., vol. 11, no. 1, pp. 73–107, 1990.
85. Z. Amjad, "Controlling metal ion fouling in industrial water systems," Ultra Pure Water, vol. 17, no. 4, 31–40.
86. Vosough A, Assari MR, Peyghambarzadeh SM, Azizi S. Influence of fluid flow rate on the fouling resistance of calcium sulfate aqueous solution in subcooled flow boiling condition. Int J Therm Sci 2020; 154: 106397.
87. Mullin, J. W., Crystallization, 3rd ed., Butterworth-Heinemann, Oxford, UK, 1993.
88. Abd-Elhady MS, Malayeri MR. Transition of convective heat transfer to subcooled flow boiling due to crystallization fouling. Appl Therm Eng 2016; 92: 122–129.
89. Malayeri MR, Müller-Steinhagen H, Bartlett TH. Fouling of tube bundles under pool boiling conditions. Chem Eng Sci 2005; 60: 1503–1513.
90. Esawy, M., 2011, Fouling of Structured Surfaces during Pool Boiling of Aqueous Solutions, Ph.D. thesis, Institute for Thermodynamics and Thermal Engineering, University of Stuttgart.
91. M. Esawy, M.S. Abd-Elhady, M.R. Malayeri and H. Müller-Steinhagen, "Influence of sintering on deposit formation during pool boiling of calcium sulphate solutions," Exp. Therm. Fluid Sci., vol. 34, no. 8, pp. 1439–1447, 2010.
92. M. Förster and M. Bohnet, "Influence of the interfacial free energy crystal/heat transfer surface on the induction period during fouling," Int. J. Therm. Sci., vol. 38, pp. 944–954, 1999.

Chapter 3
Growth Period in Crystallization Fouling

Abstract In the present chapter, recent decade's research on the crystallization fouling of calcium salts including calcium sulfate ($CaSO_4$), calcium carbonate ($CaCO_3$) and their combination were reviewed. The effects of the most important operational parameters affecting the formation of crystalline fouling such as flow velocity (u), fluid bulk temperature (T_b) and heat transfer surface temperature (T_s), salt concentration, applied heat flux (HF), and pH were categorized from various papers. The effects of these parameters were explained in the fouling growth period. Our goal is to create general ideas among crystallization fouling research and to identify the effects of various factors on this type of fouling. In addition, it is tried to show the strengths and weaknesses of the previous researches and the possibility of creating new research areas.

Keywords Crystallization fouling · Calcium salts · Fouling growth period · Operational parameters

3.1 Introduction

The steps taken when designing heat exchangers are the most effective steps in controlling fouling. These steps include selecting the appropriate type of heat exchanger, selecting the optimal operating conditions (higher fluid velocity, lower temperatures, etc.), and selecting the heat exchanger optimal design. Gilmour [1] reported that poor design of the shell side in shell and tube heat exchangers is the main cause of fouling and thus reduced heat exchanger performance. Plate heat exchangers (PHE) received more attention than shell and tube mainly due to their lower failure [2]. If the factors affecting the deposition are accurately investigated, fouling can be prevented or reduced [3]. In addition, the costs of fouling can be reduced by identifying and controlling the governing mechanisms [4]. In this chapter, the studies in which the effects of the parameters shown more clearly were selected to investigate the effect of the parameters. The main papers used in the analysis of the effect of the parameters are summarized in Table 3.1. In this chapter, recent decade's research

M. Varnaseri and S. M. Peyghambarzadeh, *Scale Formation in Heat Exchangers*,
https://doi.org/10.1007/978-3-031-52704-3_3

on the crystallization fouling of calcium salts including calcium sulfate ($CaSO_4$), calcium carbonate ($CaCO_3$) and their combination were reviewed. The effects of the most important operational parameters affecting the formation of crystalline fouling such as flow velocity, heated surface and bulk temperature, applied heat flux, salt concentration and pH were categorized from various papers. The effects of these parameters were explained in the fouling growth period. Our goal is to create general ideas among crystallization fouling research and to identify the effects of various factors on this type of fouling. In addition, it is tried to show the strengths and weaknesses of the previous researches and the possibility of creating new research areas.

3.2 Effects of Flow Velocity

Flow velocity is perhaps the most important operating parameter in controlling fouling in processes. To consider the influence of flow velocity on the fouling growth period, between many studies that have been performed, some important results were selected and presented in Figs. 3.1, 3.2, 3.3, 3.4, 3.5, 3.6, 3.7, 3.8, 3.9, 3.10, 3.11, 3.12, 3.13, 3.14 and 3.15. The selection criteria for the studies presented in Figs. 3.1, 3.2, 3.3, 3.4, 3.5, 3.6, 3.7, 3.8, 3.9, 3.10, 3.11, 3.12, 3.13, 3.14 and 3.15 are:

- Display the data in the form of fouling resistance (R_f) against time.
- Presentation of the effect of parameters on the fouling resistance clearly.
- The data have been widely cited by other researchers.

In addition to classifying studies based on the type of salt, the flow velocity effect on crystallization fouling during fouling growth period can be divided into two main groups according to the fouling mechanism. Although these studies can be divided according to the heat transfer conditions as presented in Chap. 1, due to the complexity and interference of these classifications, in the present study, priority is given to the controlling mechanism and then, to the heat transfer conditions.

A group of researchers believes that the fouling mechanism is mass transfer-controlled in laminar sub-layer. As can be seen in Fig. 3.1, increasing flow rate from 0.5 to 2 m/s, the fouling resistance also increased. These data were extracted from the study of Peyghambarzadeh et al. [57] who investigated the operating parameters effect on fouling resistance of a solution containing 2.2 g/l of $CaSO_4$ under subcooled flow boiling conditions.

Other operating conditions of the research can be seen in Table 3.1. They attributed this phenomenon to the dependence of fouling resistance on surface temperature, and stated that at low surface temperature, where the mass transfer mechanism controls the fouling process, fouling resistance increases with increasing flow velocity. Vosough et al. [4, 36] also stated that for $CaSO_4$ fouling in subcooled flow boiling conditions at low heat flux (equivalent to low surface temperature), the mass transfer mechanism is dominant, and the fouling resistance increases with increasing flow velocity (or Reynolds number). Similar results to Fig. 3.1 have been observed in

Table 3.1 Crystallization fouling studies and their operating conditions

References	System	Heat transfer regime	Operating conditions
$CaSO_4$			
Hasson and Zahavi [5]	Annular heat exchanger	FC	C_b: 2010–2030 ppm, T_b: 50–55 °C, T_s: 80–87 °C, u: 0.165–0.223 m/s
Jamialahmadi et al. [6]	Boiling vessel with a horizontal heater	PB	C_b: 1.6 g/l, T_b: 100 °C, $\dot{q}$: 9.631–77.045 kW/m^2
Najibi et al. [7]	Vertical annulus	SFB	C_b: 1.6–2.7 g/l, T_b: 65–95 °C, T_s: 95–140 °C, u: 0.5–2 m/s
Müller-Steinhagen [8]	PHE	FC	C_b: 2988–3198 ppm, $t_{h,in}$: 76–90 °C, $t_{s,in}$: 41–52 °C, u: 0.22–0.73 m/s
Förster et al. [9]	Annular test tube and PHE	FC	C_b: 2.5 g/l, $t_{h,in}$: 75 °C, $\dot{q}$: 31.8 kW/m^2, u: 0.2–1 m/s
Müller-Steinhagen et al. [10]	Vertical annulus	FC and SFB	C_b: 2–2.5 g/l, T_b: 80 °C, u: 0.4–1.5 m/s, $\dot{q}$: 100–300 kW/m^2
Bansal et al. [11]	PHE	FC (fully turbulent)	C_b: 2.825–3.196 kg/m^3, $t_{s,in}$: 40–55 °C, $t_{h,in}$: 77.5–93 °C, u: 0.18–0.93 m/s
Jamialahmadi et al. [12]	Boiling vessel with a horizontal heater	PB	C_b: 0.8–1.6 g/l, T_b: 100 °C, T_s: 100–125 °C, $\dot{q}$: 10–300 kW/m^2
Jamialahmadi and Müller-Steinhagen [13]	Boiling vessel with a horizontal heater	PB	C_b: 0.8–2 kg/m^3, T_b: 100 °C, $\dot{q}$: 35–480 kW/m^2
Malayeri et al. [14]	Boiling vessel with a horizontal heater	PB	C_b: 0.8–1.6 g/l, T_b: 100 °C, $\dot{q}$: 33–300 kW/m^2
Behbahani et al. [15]	Vertical annulus	FC	C_b: 1.4–1.85%wt., T_b: 73–78 °C, T_s: 102–129 °C, u: 1.3–1.8 m/s
Mwaba et al. [16]	Horizontal annulus	FC (fully turbulent)	$C_{b,init}$: 3 kg/m^3, T_b: 40 °C, T_w: 65–82 °C, Re: 11,000–34,000
Malayeri and Müller-Steinhagen [17]	Boiling vessel with a horizontal heater	PB	C_b: 0.8–1.6 g/l, T_b: 100 °C, T_s: 104.3–114.1 °C, $\dot{q}$: 55–300 kW/m^2
Fahiminia et al. [18]	Vertical annulus	FC (fully turbulent)	C_b: 3100–3600 ppm, T_b: 51–63 °C, T_w: 66–83 °C, u: 0.1–1.6 m/s, Re: 2100–36,000
Herz et al. [19]	Vertical rectangular duct	FC	C_b: 3.75–4.4 g/l, T_b: 40 °C, T_s: 69–82 °C, u: 0.15 m/s, $\dot{q}$: 80–120 kW/m^2

(continued)

Table 3.1 (continued)

References	System	Heat transfer regime	Operating conditions
Alahmad [20]	Horizontal annulus	FC	C_b: 500–1500 ppm, T_b: 40–60 °C, u: 0.05–0.37 m/s, Re: 400–29,600, pH: 4–9
Al-Janabi et al. [21–23]	Vertical rectangular duct	FC	C_b: 4 g/l, T_b: 40 °C, T_s: 60–82 °C, u: 0.1–0.3 m/s, Re: 4480, $\dot{q}$: 120 kW/m^2
Esawy et al. [24]	Boiling vessel with a horizontal heater	PB	C_b: 1.6 mg/ml, T_b: 100 °C, $T_{s,\ inner}$: 100–211 °C, $\dot{q}$: 40–300 kW/m^2
Bansal et al. [25]	Plate and double-pipe heat exchanger	FC	C_b: 2.9727–2.9850 kg/m^3, $t_{h,in}$: 86–90 °C, $t_{s,in}$: 40–65 °C, u: 0.182–0.705 m/s
Esawy et al. [26]	Cylindrical boiling vessel	PB	C_b: 1.2–1.6 g/l, T_b: 101 °C, T_s: max 170 °C, $\dot{q}$: 5–300 kW/m^2
Peyghambarzadeh et al. [27]	Vertical annulus	SFB	C_b: 1.5–2.2 g/l, T_w: 102–115 °C, $\dot{q}$: 400 kW/m^2, u: 0.5–2 m/s
Peyghambarzadeh et al. [28]	Vertical annulus	SFB and FC	C_b: 0–0.2 kg/m^3, T_b: 300–363 K, T_w: 300–390 K, $\dot{q}$: 0–100 kW/m^2, u: 0.1–2 m/s
Peyghambarzadeh and Bahrami [29]	Vertical annulus	SFB	C_b: 1.5–1.8 mg/ml, T_b: 85–93 °C, u: 0.8–1 m/s, $\dot{q}$: 300–350 kW/m^2
Lee et al. [30]	PHE	FC	C_b: 0.40–0.66%wt., $t_{s,in}$: 15–55 °C, $t_{h,in}$: 55–95 °C, Re: 2500–5600
Abd-Elhady et al. [31]	Tubular heat exchangers (with injection projectiles system)	FC	C_b: 4.6 g/l, T_b: 40–80 °C, T_s: 135 °C, u: 0.5–3.5 m/s, $\dot{q}_{max}$: 570 kW/m^2
Abd-Elhady and Malayeri [32]	Annulus	FC and FB	C_b: 3–4 g/l, T_b: 40 °C, u: 0.8 m/s, T_s: 80–240 °C
Han et al. [33]	Half cylinder vortex generators in a rectangular channel	FC	C_b: 2–4 kg/m^3, T_w: 320–340 K, T_b: 300 K, u: 0.4–0.8 m/s
Maddahi et al. [34, 35]	Heat exchanger (liquid–solid fluidized bed)	FC	C_b: 1.2–1.9 kg/m^3, T_b: 65 °C, T_w: 67–75 °C, u: 0.15–0.52 m/s, $\dot{q}$: 25–82 kW/m^2
Vosough et al. [4]	Vertical annulus	SFB	C_b: 1.7–2.2 mg/ml, T_b: 55–75 °C, $\dot{q}$: 8–95 kW/m^2, Q: 2.5–11.5 l/min
Vosough et al. [36]	Vertical upward annulus	SFB	C_b: 1.75–2.20 g/l, $\Delta T_{fluid\ subcooled}$: 24–45 °C, $\dot{q}$: 8–95 kW/m^2, Q: 2.5–5.5 l/min, Re: 5000–15,000

(continued)

Table 3.1 (continued)

References	System	Heat transfer regime	Operating conditions
$CaCO_3$			
Hasson et al. [37]	Constant heat-flux annular exchanger	FC	C_b: 300–320 ppm, T_b: 45 ± 0.8 °C, T_s: 67–85 °C, u: 0.25–0.82 m/s, Re: 13,000–42,000, pH: 7.35–7.5
Sheikholeslami and Watkinson [38]	Plain and externally finned surface (annulus)	FC	C_b: 4.5–20 mg/l, u: 0.3–0.8 m/s, q: 8.4–15.45 kW, Re: 7000–29,000, pH: 7.45–7.93
Najibi et al. [39]	Vertical annulus	SFB	C_b: 0.0035–0.00 75 mol/l, T_b: 80 °C, T_s: 105–120 °C, u: 0.6–1.8 m/s, pH: 7.2–7.7
Yang et al. [40, 41]	Boiling vessel with horizontal heater	PB	[$CaCl_2$]: 0.49 g/l, [$NaHCO_3$]: 0.75 g/l, T_b: 41 °C, T_s: 73.5–78.7 °C, u: 1.28–1.62 m/s, $\dot{q}$: 110–463 kW/m^2
Xiaokai et al. [42]	Vertical annulus	SFB	C_b: 3–10 mmol/l, T_b: 71–88 °C, T_s: 95–135 °C, u: 0.3–0.9 m/s, $\dot{q}$: 110–250 kW/m^2
Zhenhua et al. [43]	Horizontal double-pipe heat exchanger	FC	C_b: 300–750 mg/l, $t_{h,in}$: 58–90 °C, $t_{s,in}$: 22.5–44.5 °C, u: 0.6–1.8 m/s
Pääkkönen et al. [44, 45]	PHE	FC	C_b: 0.0034 mol/l, T_b: 30–40 °C, T_w: 75–95 °C, u: 0.2–0.4 m/s, $\dot{q}$: 49.3–61.8 kW/m^2
Tijing et al. [46]	Double-pipe heat exchanger	FC	C_b: 350–550 ppm, $t_{h,in}$: 85.5 ± 0.5 °C, $t_{s,in}$: 23.5 ± 0.5 °C, u_h: 1.1–1.3 m/s, u_s: 0.3–1 m/s
Al-Hadhrami et al. [47]	Twisted tube heat exchanger	FC	C_b: 0.15 mg/ml, T_b: 43.3–75.4 °C, u: 0.5–2 m/s, q: 200–800 W
Cai et al. [48]		PB	C_b: 65–750 mg/l, $\dot{q}$: 2.71–70.05 kW/m^2
Wang et al. [49]	Double-pipe heat exchanger	FC	C_b: 375 mg/l, $t_{s,in}$: 47.5 °C, $t_{h,in}$: 80 °C, u: 0.06–0.8 m/s, Re: 618–9406, $\dot{q}$: 3.9–52.1 kW/m^2
Teng et al. [50]	Double-pipe heat exchanger	FC	C_b: 300–500 mg/l, T_b: 50–70 °C, u: 0.15–0.45 m/s
Al-Gailani et al. [51]	Flow cell (flow is uniform with no fluid recirculation)	FC (laminar flow)	C_b: 3.07 mmol/l, T_s: 90 °C, Q: 0.65–6.32 ml/min, Re: 1.26–12.07, pH: 7.2
$CaSO_4$/$CaCO_3$			
Helalizadeh et al. [52, 53]	Vertical annulus	FC and SFB	[$CaSO_4$]: 1–2.5 g/l, [$CaCO_3$]: 0.25–1 g/l, T_b: 50–90 °C, T_s: 103–117 °C, u: 0.5–2 m/s, $\dot{q}$: 100–400 kW/m^2

(continued)

Table 3.1 (continued)

References	System	Heat transfer regime	Operating conditions
Helalizadeh et al. [54]	Two parallel annular test heaters	FC and SFB	$[CaSO_4]$: 1–2.5 g/l, $[CaCO_3]$: 0–0.03 g/l, T_b: 50–90 °C, T_s: 103–119 °C, u: 0.5–2 m/s, $\dot{q}$: 100–450 kW/m^2
Sudmalis and Sheikholeslami [55]	–	–	$[CaSO_4]$: 0.005–0.02 M, $[CaCO_3]$: 0.005–0.02 M, T_b: 60–80 °C
Song et al. [56]	PHE	FC	$[CaSO_4]$: 2–4 kg/m^3, $[CaCO_3]$: 1–2 kg/m^3, $t_{s,in}$: 15 °C, $t_{h,in}$: 55–95 °C, u: 0.5–2 m/s, $\dot{q}$: 100–400 kW/m^2, Re: 1800–3900, pH: 6.9–7.1

FC forced convection; *PB* pool boiling; *SFB* subcooled flow boiling; *PHE* plate heat exchanger; C_b bulk concentration; T_b bulk temperature; T_s surface temperature; u flow velocity; $\dot{q}$ heat flux; $t_{h,in}$ inlet hot temperature; $t_{s,in}$ inlet solution temperature; $C_{b,init}$ initial bulk concentration; T_w wall temperature; $T_{s,\ inner}$ inner surface temperature

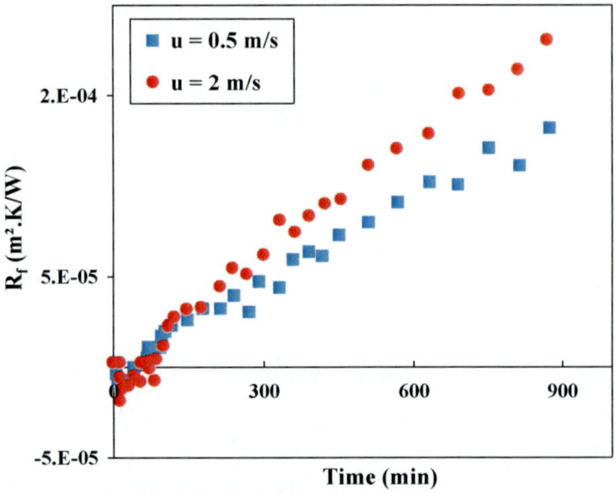

Fig. 3.1 Flow velocity effect on $CaSO_4$ fouling resistance (adapted from [57])

the studies of Najibi et al. [7] and Helalizadeh et al. [52] for $CaSO_4$ salt. In fact, if process is diffusion-controlled, higher flow velocity causes higher fouling resistance [15].

Figures 3.2 and 3.3 show the flow velocity effect in 0.6–1.4 m/s range of under subcooled flow boiling conditions for $CaSO_4$ salt at a concentration of 2 g/l [7]. Other operating conditions of the research are available in Table 3.1. As can be seen, at all velocities studied, after an initial period where the effects of surface nucleation are predominant, over time fouling resistance increases linearly. Najibi et al. [7] stated that mass transfer boundary layer thickness is large at low velocities, and as a result, fouling is more affected by molecular diffusion.

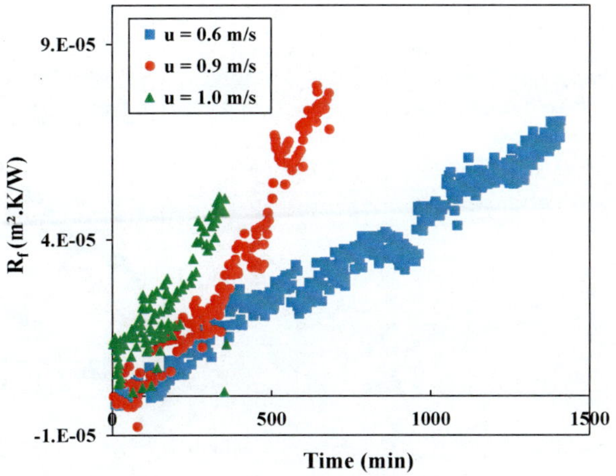

Fig. 3.2 Flow velocity effect on $CaSO_4$ fouling resistance (T_S: 111 °C) (adapted from [7])

As the flow velocity increases, the boundary layer becomes thinner, thus reducing the molecular diffusion effect compared to the surface chemical reaction. During the subcooled flow boiling condition, depending on the surface temperature or heat flux, the growing vapor bubbles cover parts of heated surface. It can be said that due to the disturbance of the liquid layer near the surface by the growing bubbles the fouling process near these bubbles is controlled by reaction. As can be concluded from Figs. 3.2 and 3.3 in subcooled flow boiling conditions, increase of fouling resistance occurred with increasing flow velocity, therefore, the fouling mechanism is controlled by mass transfer. Furthermore, the fouling curves in Figs. 3.2 and 3.3 are linear, as a result, the removal process is negligible and the fouling rate is constant. On the other hand, groups of researchers have stated that the crystallization fouling mechanism is chemical reaction-controlled, and therefore, independent of the flow velocity.

In Figs. 3.4 and 3.5, it is clear that $CaSO_4$ fouling resistance decreased with increasing flow velocity. These data were obtained under the forced convection conditions. Figure 3.4 reported by Krause [58] at two flow velocities of 0.5 and 1.33 m/s for crystallization fouling of $CaSO_4$ salt. Figure 3.5 shows $CaSO_4$ deposition results in a plate heat exchanger (PHE) investigated by Müller-Steinhagen [8] and Bansal et al. [2]. Operating conditions can be seen in Table 3.1. There was no filter in the apparatus system and the studied velocities were in 0.24–0.67 m/s range. It was reported that the fouling resistance decreased with increasing the flow velocity due to the increase in shear stress and increase of the removal forces.

Hasson [59] presented a model for predicting scale formation and stated that for flow velocities below 0.8 m/s, the fouling resistance increased with increasing fluid velocity. This phenomenon occurs due to the rapid reaction between the ions at heated surface, and shifts the rate determination phase from the diffusion in the bulk flow to the interface of the solid/fluid. He stated that at flow velocity higher than 0.8 m/s, asymptotic fouling resistance decreased with increasing flow velocity.

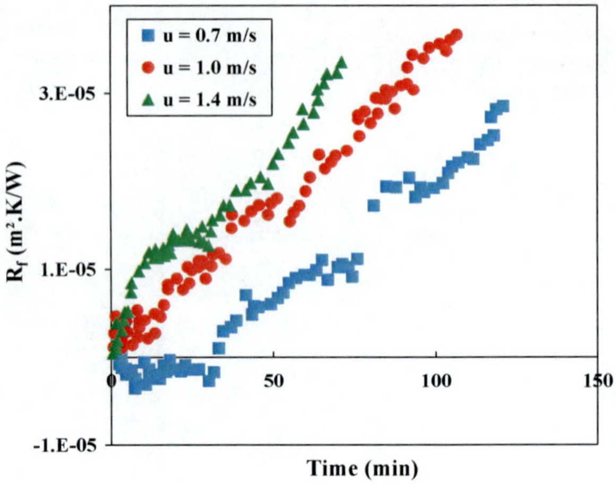

Fig. 3.3 Flow velocity effect on $CaSO_4$ fouling resistance (T_S: 115 °C) (adapted from [7])

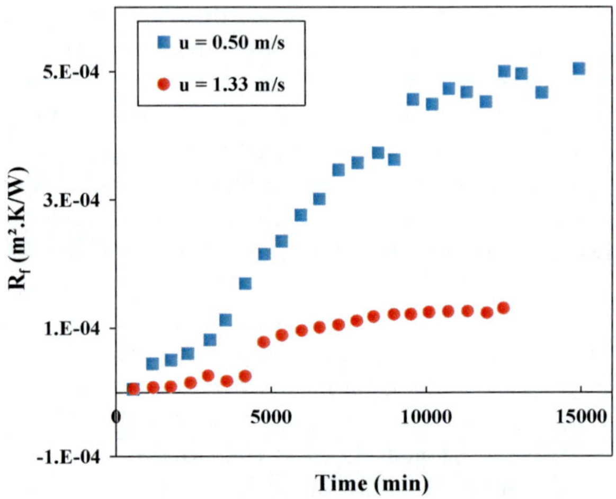

Fig. 3.4 Flow velocity effect on $CaSO_4$ fouling resistance (adapted from [58])

By increasing the flow velocity, due to the increase in flow turbulence, the rate of heat transfer from the surface increases and as a result, the surface temperature decreases. Then, this temperature affects the fouling reaction rate constant according to Eq. (2.1).

Han et al. [33] investigated the fouling properties of $CaSO_4$ in rectangular channels with vortex generators and Fig. 3.6 shows the results of their work. Table 3.1 presents the operating conditions of the research. They stated that since the fouling resistance is the result of deposition and removal rates, as the flow velocity increases, a greater shear force and flushing occur, and thus, the removal rate exceeds the deposition rate and reduces the fouling resistance.

Förster et al. [9] reported some results from experiments in 0.5–1 m/s flow velocity range for $CaSO_4$ scale formation. They showed a known flow velocity effect on

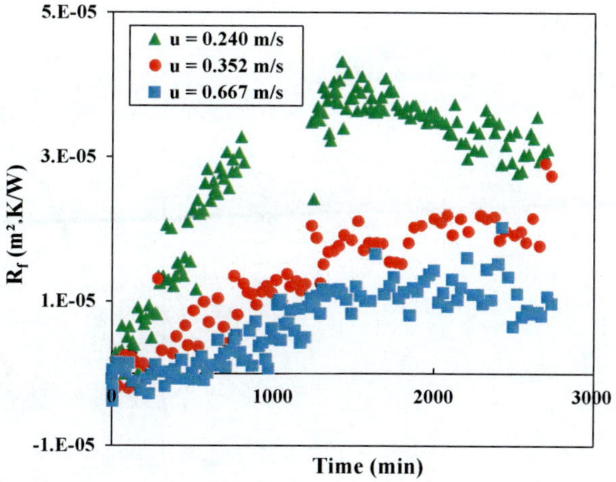

Fig. 3.5 Flow velocity effect on $CaSO_4$ fouling resistance (adapted from [2, 8])

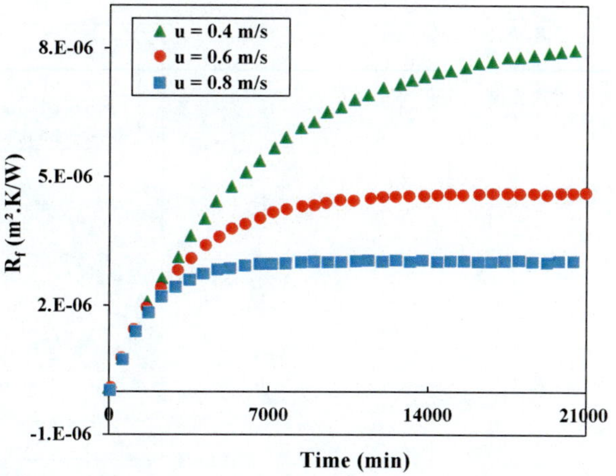

Fig. 3.6 Flow velocity effect on $CaSO_4$ fouling resistance (adapted from [33])

fouling, i.e., fouling resistance decreased with increasing flow velocity (Fig. 3.7). Ritter [60] previously also stated this point.

Figure 3.8 represent the results of various research on flow velocity effect on fouling resistance as a function of time for $CaCO_3$ salt. Contradicting observations were also reported about the influence of flow velocity on $CaCO_3$ fouling similar to those reported for $CaSO_4$. As shown in Figs. 3.8 and 3.9, by increasing velocity the fouling resistance increased. Figure 3.8 clearly shows that at all studied velocities after an initial period; there is a linear increase in fouling resistance over time. Figure 3.8 is obtained under subcooled flow boiling conditions for $CaCO_3$ salt in the flow velocity range of 0.9–1.6 m/s [39]. Table 3.1 presents other operational parameters.

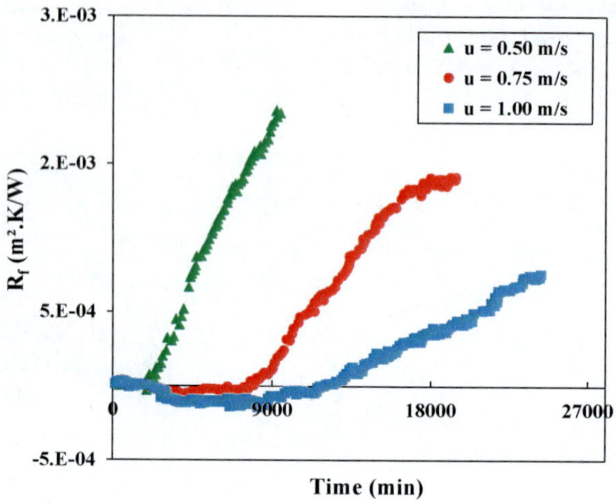

Fig. 3.7 Flow velocity effect on $CaSO_4$ fouling resistance (adapted from [9])

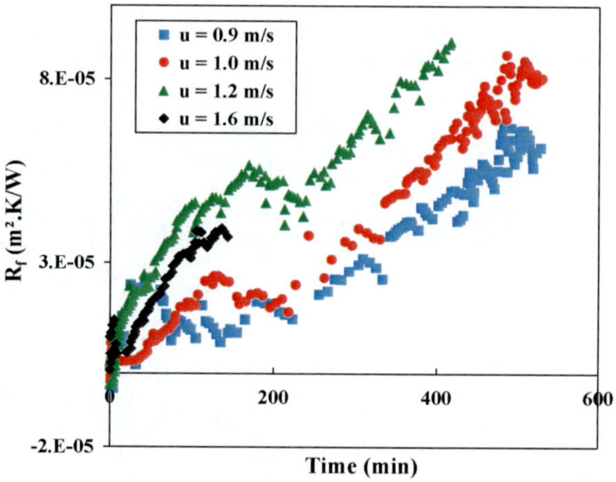

Fig. 3.8 Flow velocity effect on $CaCO_3$ fouling resistance (adapted from [39])

In Fig. 3.8, because the fouling process is controlled by mass transfer, fouling resistance increased with increasing flow velocity. This result is in Accordance with Hatch [61] who stated that increasing the flow velocity accelerates the fouling resistance. However, a remarkable point in Fig. 3.8 is decrease in fouling resistance at u = 1.6 m/s compared to u = 1.2 m/s, which can be due to the change of fouling mechanism from mass transfer-controlled to surface reaction-controlled.

Xiaokai et al. [42] studied the $CaCO_3$ fouling mechanism under subcooled flow boiling conditions at two different concentrations. Table 3.1 presents the operational variables of the experiments. One of the results of this research is presented in Fig. 3.9. At this concentration of salt ($[NaHCO_3]$ = 20 mmol/l, $[CaCl_2]$ = 10 mmol/l), the nucleation rate was relatively high, and the heat transfer surface was quickly covered by $CaCO_3$ fouling, so no induction period was observed. The effect of flow velocity in

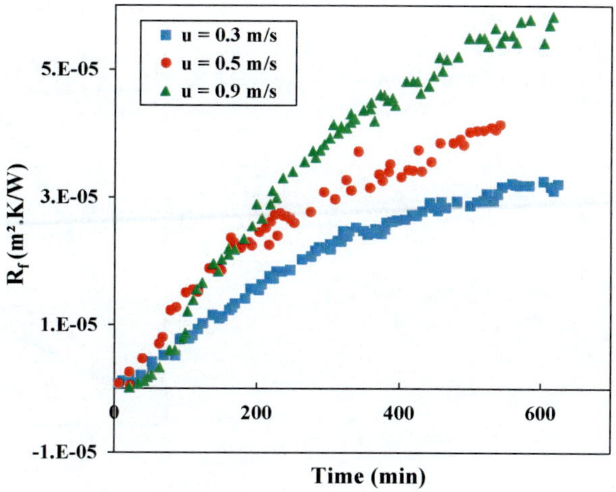

Fig. 3.9 Flow velocity effect on $CaCO_3$ fouling resistance (adapted from [42])

these conditions was very severe, and due to diffusion-controlled fouling, as the flow velocity increased fouling resistance increased, too. In other experiments at lower concentrations, not shown here ($[NaHCO_3] = 10$ mmol/l, $[CaCl_2] = 5$ mmol/l), an initial period in fouling process was observed. At this salt concentration, because the "surface reaction" controlled the initial growth of the nuclei/crystal, the crystals growth and surface coating were relatively slow, and flow velocity had almost no effect on fouling resistance.

In Figs. 3.10, 3.11, 3.12, 3.13 and 3.14, it is clear that with increasing fluid velocity, the fouling resistance decreased. On the other hand, unlike Figs. 3.1, 3.2, 3.3, 3.4, 3.5, 3.6, 3.7, 3.8 and 3.9, where the mechanism was subcooled flow boiling and the curves were almost linear, in these figures, heat transfer mechanism is forced convection, and the curves are asymptotic. In addition, these results are very similar to the findings of Yu et al. [62] who concluded that increasing the fluid velocity increases the removal rate due to the increase in drag and shear stress in tube wall. It is also in agreement with the results of Watkinson and Martinez [63] who reported that fouling resistance decreases with increasing fluid velocity. Contrary to the findings of Figs. 3.8 and 3.9, Walker and Sheikholeslami [64] argued in their research that fouling thickness decreases with increasing flow velocity in fully developed turbulent flow. Other researchers [20, 42, 50, 65, 66] also expressed the similar results. In Fig. 3.10, the fouling resistance curves, after a linear increase at all studied flow velocity, finally reached almost constant values [67, 68].

According to Fig. 3.10, the fluid velocity has a clear effect on fouling resistance with tripling the fluid velocity (0.6–1.8 m/s), the fouling resistance decreased to about 20% of its initial value. Thus, surface integration reaction controls the experiments, and the shear stress due to the increased flow velocity is the main obstacle to the adherence of deposit on the surface. Zhenhua et al. [43] obtained these results for $CaCO_3$ under forced convection conditions in a smooth tube.

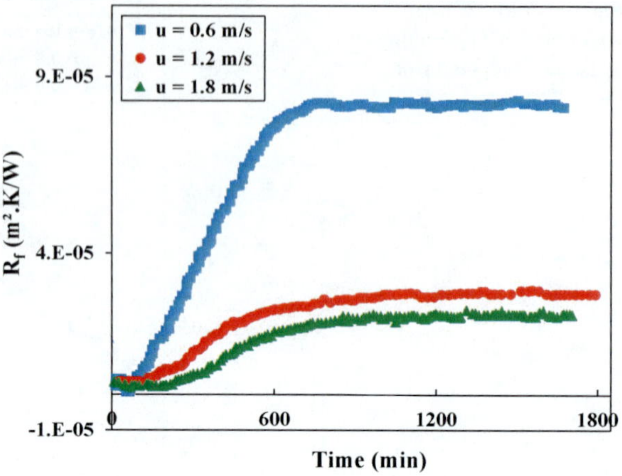

Fig. 3.10 Flow velocity effect on $CaCO_3$ fouling resistance (adapted from [43])

Al-Hadhrami et al. [47] presented Fig. 3.11 by examining $CaCO_3$ fouling in a twisted tube heat exchanger, the fluid velocity was changed in the range of 0.5–2 m/s. Figure 3.11 shows similar results to Figs. 3.10, 3.12 and 3.13. The remarkable point of Fig. 3.11 is that the fouling resistance maximum decreases with flow velocity increasing at the initial 1000 min of the experiments. This behavior can be justified by the fact that at high flow velocities, the amount of flow turbulence intensifies and causes deposit particles to be removed. At higher flow velocities, growth of fouling layer continues until it reaches a critical thickness, as a result, fouling resistance also increases. This critical thickness is achieved much faster at low flow velocity and then there is an asymptotic trend. Schlichting [69] showed that as the flow velocity increases, the viscous sub-layer becomes thinner and the wall temperature decreases thus the fouling rate is reduced.

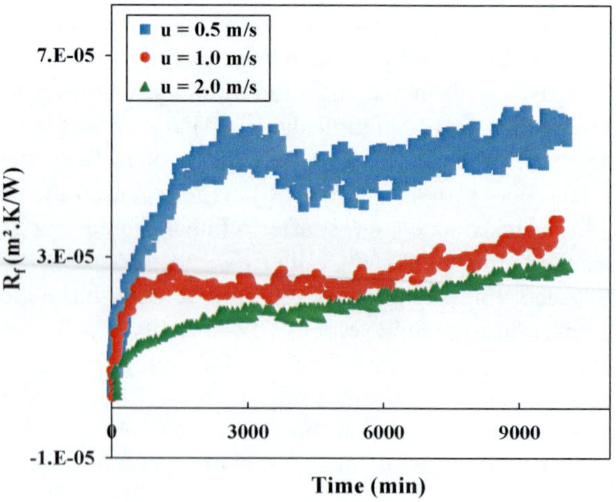

Fig. 3.11 Flow velocity effect on $CaCO_3$ fouling resistance (adapted from [47])

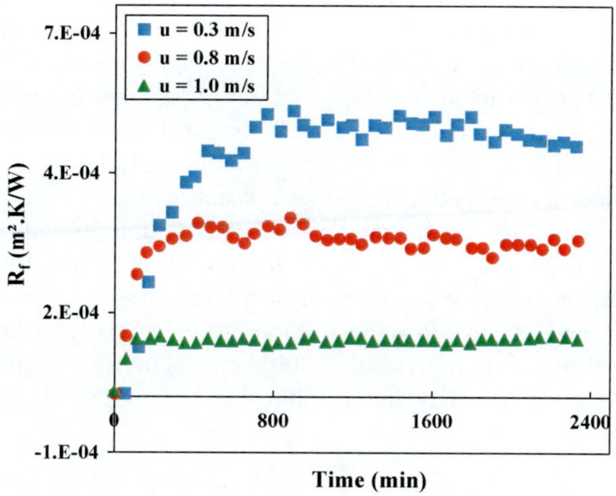

Fig. 3.12 Flow velocity effect on $CaCO_3$ fouling resistance (adapted from [46])

Figure 3.12 was taken from the research of Tijing et al. [46] on the use of radio frequency electric field to remove $CaCO_3$ fouling from a copper tube. The heat transfer condition was forced convection and only the flow velocity in cold-side water was changed. Other operating conditions are presented in Table 3.1. The results showed a decrease in fouling resistance with increasing cooling water flow velocity. At the lowest u = 0.3 m/s, the value of asymptotic fouling resistance is at its highest value (4.46×10^{-4} m^2 K/W). At this u = 0.3 m/s, compared to other flow velocities, the shear stress was at its lowest level and was not able to remove fouling. For higher velocity such as 0.8–1 m/s, the asymptomatic fouling resistance value decreased to about 2.51×10^{-4} m^2 K/W and 1.01×10^{-4} m^2 K/W, respectively. Therefore, they concluded that at higher flow velocity, even if fouling resistance is higher, fluid flow shear stresses are large enough to remove fouling continuously.

In Fig. 3.13 by keeping the surface integration rate constant by keeping the initial wall temperature constant (T_s = 85 °C), and flow velocity is varied from 0.2 to 0.4 m/s to determine the mechanism of $CaCO_3$ fouling process. Since the test results were presented in a relatively short time (about 350 min), asymptotic fouling behavior was not observed, therefore, the fouling growth period is in its primary step. In the initial fouling growth period, the number of crystals is low and the removal rate is negligible due to the small effect of these crystals on the cross-sectional area of the flow channel [44, 45, 66, 70]. In addition, $CaCO_3$ deposition is relatively strong, and difficult to remove [71]. Thus, a linear fouling resistance curve was observed. However, in Fig. 3.12 with increasing test time (up to about 1800 min) it seems that crystals became longer and more brittle and effect of local shear stress caused the fouling resistance curve to have an asymptotic shape [72]. Figure 3.13 shows that surface integration overcame mass transfer [8, 66, 72, 73]. By increasing the fluid velocity, shear stress at fluid–solid interface prevents the crystals from adhering to crystal lattice. Increasing flow velocity also reduces the fluid residence time near the heated surface and thus, reduces the possibility of foulant adhering to the surface.

Wang et al. [49] investigated the flow velocity effect on $CaCO_3$ deposition properties in a smooth tube of a double-pipe heat exchanger. Figure 3.14 shows the results of this experiment. They showed four time zones in their test results.

The first region is positive fouling resistance, the second region is negative fouling resistance, the third region is continuously increasing fouling resistance, and the fourth region is where fouling resistance remains constant. With increasing Reynolds number (in the range of 618–2100), fouling growth rate increases until it reaches a maximum value, then decreases with increasing Re number. This may be due to changes in flow properties near the wall. Fouling growth rate increases under laminar flow is increased due to thick viscous sublayer. Under turbulent flow conditions (Reynolds number above 3000), removal rate was significantly increased and fouling growth rate was no longer influenced by flow velocity. In general, with increasing flow

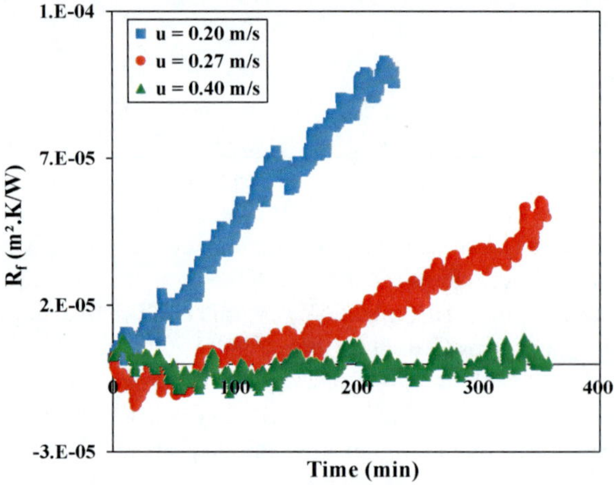

Fig. 3.13 Flow velocity effect on $CaCO_3$ fouling resistance (adapted from [44, 45])

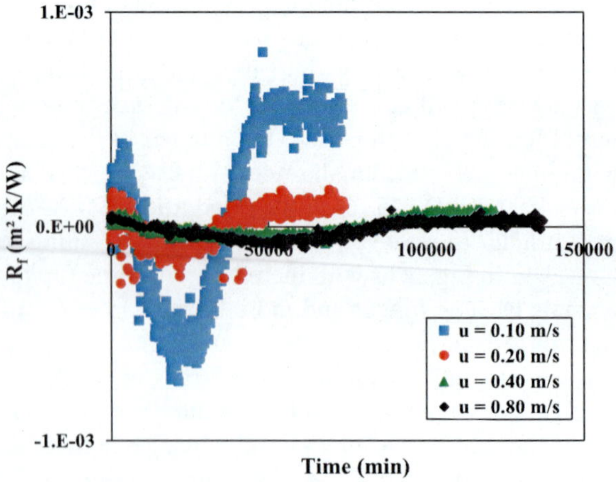

Fig. 3.14 Flow velocity effect on $CaCO_3$ fouling resistance (adapted from [49])

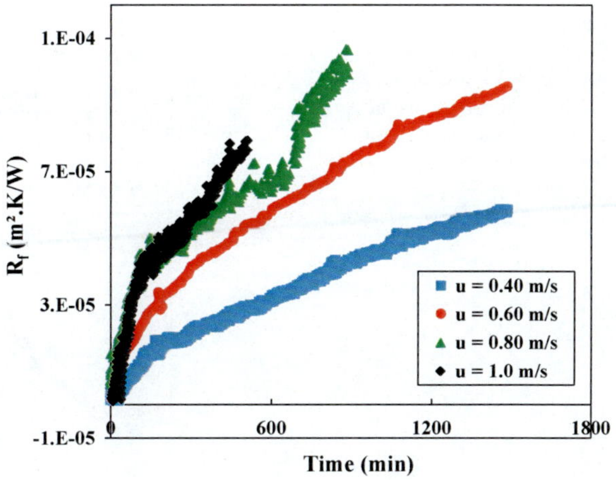

Fig. 3.15 Flow velocity effect on mixed salts fouling resistance (adapted from [52])

velocity, if fouling process is mass transfer-controlled, fouling resistance increases and if fouling process is surface integration-controlled, fouling resistance decreases [3, 18, 74].

Xiaokai et al. [42] stated that in deposition such as $CaCO_3/CaSO_4$, which is relatively strong deposition, increasing the flow velocity mat have no effect on fouling reduction. Helalizadeh et al. [52] studied the crystallization fouling mechanisms of mixed salt ($CaCO_3$ and $CaSO_4$) under subcooled flow boiling and convective heat transfer. The results of flow velocity effect on fouling resistance of this investigate are shown in Fig. 3.15, flow velocity varied of 0.4–1 m/s were applied. For all velocities studied, fouling resistance after a rapid deposition period increases linearly with time. The fouling process at low flow velocities was clearly controlled by diffusion.

As previously discussed about the velocity effect on fouling resistance, in Helalizadeh et al.'s [52] research, increasing velocity reduces the boundary layer thickness and thus changes the mechanism controlling the fouling process from molecular diffusion to reaction. In addition, they showed that the effect of increasing the Reynolds number was to increase the removal rate, which also reduced the asymptotic fouling resistance.

The comparison results of Peyghambarzadeh et al. [28] in field of the heat transfer mechanism effect as well as the bubble growth on $CaSO_4$ crystallization fouling is presented in Fig. 3.16. Fouling was formed rapidly on the surface under subcooled flow boiling conditions ($T_b = 80$ °C) compared to the single-phase convective conditions ($T_b = 40$ °C). Figure 3.16 shows that the bubble activity in subcooled flow boiling conditions compared to force convective caused an intense increase in the fouling resistance.

The flow velocity effect on crystallization fouling resistance depends entirely on fouling control mechanism. It can be concluded that fouling process depends on mass transfer of precursors and a formation/attachment reaction to/on heated surface. Mass transfer coefficient values and reaction rate constant determine the controlling

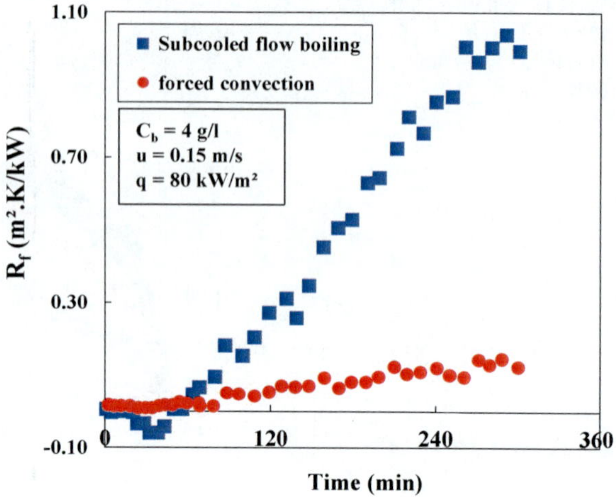

Fig. 3.16 Comparison of $CaSO_4$ crystallization fouling under subcooled flow boiling and convective heat transfer conditions (adapted from [28])

mechanism of fouling process. Mass transfer coefficient depends on fluid velocity. When mass transfer is slower than the formation/attachment reaction, flow velocity affects the fouling rate. If the formation/attachment reaction is slower than the mass transfer process, the deposition depends on the other factors such as the surface temperature. In addition, if the Reynolds number increases, the fouling controlled by mass transfer changes to the fouling controlled by reaction. On the other hand, it can be said that even if ion transfer rate to heated surface increases with increasing flow velocity, due to overcoming effects of flow shear stress on mass transfer, fouling resistance decreases. Therefore, fouling is controlled by surface integration reaction, but shear stress due to increase in fluid flow velocity is the main obstacle in preventing the deposition material from adhering to the surface.

3.3 Effect of Bulk and Surface Temperature

Fouling resistance is strongly affected by surface and bulk temperature; the main reason is an inverse dependence of the supersaturation of these soluble salts with temperature [75]. Various opinions on temperature effect on fouling have been reported [76]. This review paper implies "positive, negative or no effect" of temperature on fouling resistance. Here, in addition to the division of studies based on the type of the salts, two other divisions have been applied: the division based on fouling process controlling mechanism and division based on heat transfer conditions.

In most of the reviewed studies, it is clearly stated that increasing heated surface temperature increases the crystallization fouling resistance. However, increasing bulk temperature has less effect on crystallization fouling [7, 8, 11, 15, 28, 30, 33, 39, 57, 77]. In this section, first, the temperature effect on $CaSO_4$ crystallization fouling in various studies under subcooled flow boiling and forced convection heat transfer is reviewed, and then, the results for $CaCO_3$ fouling are presented.

Peyghambarzadeh et al. [28] studied bubble formation effect on $CaSO_4$ crystallization fouling in heat exchangers. The bulk temperature effect on fouling resistance is shown in Fig. 3.17. The experiment conditions of the research are presented in Table 3.1. The reason for choosing these two temperatures (T_b = 40 and 80 °C) was to compare and investigate the fouling resistance during subcooled flow boiling and forced convection conditions. As shown in Fig. 3.17, at subcooled flow boiling much more fouling was formed in compare with forced convection. The presence of bubbles is reason of this event. The role of bubbles produced at nucleation sites under subcooled flow boiling conditions is to increase supersaturation by increasing the concentration of salts in boundary layer. Hence, more deposition rate was obtained during subcooled flow boiling. In addition, Peyghambarzadeh et al. [57] stated that the proportion of temperature to diffusion and reaction constant is direct and exponentially, respectively. Yang et al. [40] have also confirmed this result.

Najibi et al. [7] investigated bulk temperature effect on $CaSO_4$ fouling formation during subcooled flow boiling at constant T_s = 110 °C. The result of this study is presented in Fig. 3.18. Other operating conditions of the research can be seen in Table 3.1. The experiment was performed at T_b = 70–90 °C. As shown in Fig. 3.18, the fouling resistance was almost independent of the bulk temperature. They stated that at velocities below 0.9 m/s mass transfer controls fouling process while at higher velocities (above 0.9 m/s) reaction controls fouling process.

They also investigated the effect of surface temperature in their study, which is shown in Fig. 3.19. According to these results, they stated that surface temperature had a severe effect on fouling resistance. They stated that similar to Hasson et al. [37], $CaSO_4$ deposition is controlled by mass transfer and therefore, increases linearly with the surface temperature. As the surface temperature increases, the fouling curve slope increases sharply as shown in Fig. 3.19.

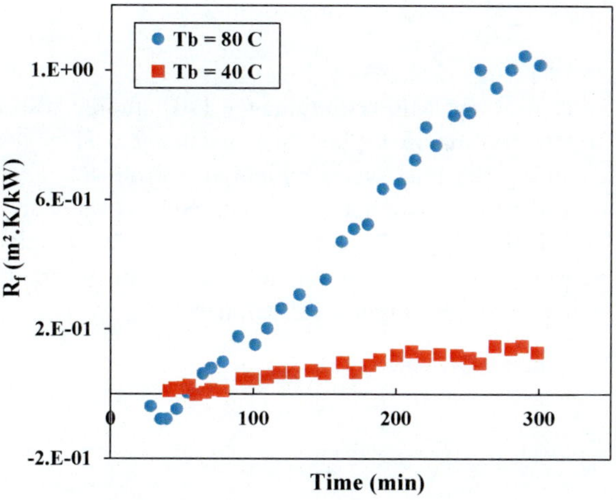

Fig. 3.17 Bulk temperature effect on $CaSO_4$ fouling resistance (adapted from [28])

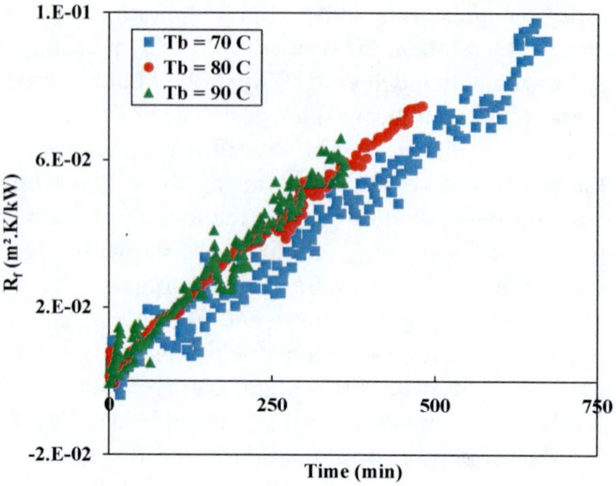

Fig. 3.18 Bulk temperature effect on $CaSO_4$ fouling resistance (adapted from [7])

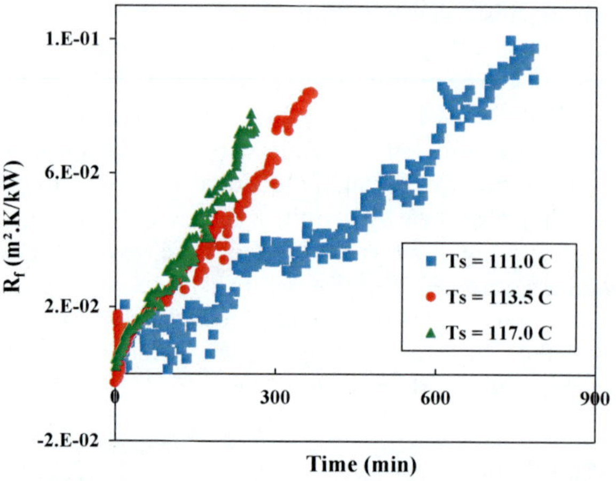

Fig. 3.19 Surface temperature effect on $CaSO_4$ fouling resistance (adapted from [7])

Bansal and Müller-Steinhagen [70] studied wall temperature effect on calcium sulfate fouling in a plate heat exchanger. They tested the fouling resistance by changing the inlet water temperature from 85 to 90 °C. The heat transfer condition of this experiment was forced convection. The results of this study, which are presented in Fig. 3.20, unlike the other studies results showed an increase in fouling resistance by increasing the inlet water temperature by only 5 °C. More fouling rate is obtained from higher wall temperatures.

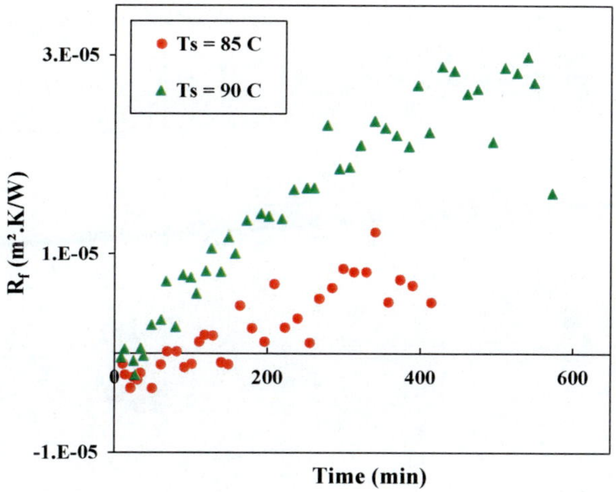

Fig. 3.20 Inlet water temperature effect on $CaSO_4$ fouling resistance (adapted from [70])

By visually observing of heat exchanger plates, the effect of wall temperature can be seen. At the hot end of the plates and the points of contact between them due to low flow velocity the wall temperature is higher and the removal rates is lower, resulting in more deposits. In addition, Müller-Steinhagen [8] achieved similar results under similar conditions and heat exchanger. These results are also confirmed in study of Bansal et al. [11].

Maddahi et al. [35] studied several parameters effect on calcium sulfate deposition along a liquid–solid fluidized bed heat exchanger with cylindrical particles. They kept bed voidage and salt concentration constant at 0.65 and 1.6 kg/m^3, respectively. Figure 3.21 show their results regarding wall temperature effect on fouling resistance. It is clear that due to the inverse solubility of $CaSO_4$, its solubility decreased with increasing temperature and more deposition was formed with increasing temperature.

At $T_w = 67.1$ °C, the fouling resistance was essentially zero and did not change with time because little deposition formed at this wall temperature. This study was performed under forced convection conditions. In addition, dominant surface temperature effect indicates that fouling process is chemical reaction-controlled.

Lee et al. [30] investigated the $CaSO_4$ crystallization fouling in a plate heat exchanger. One of the results of this study is presented in Fig. 3.22. This study was also performed under forced convection conditions. The results showed that as inlet solution temperatures ($t_{s,in}$) and inlet hot water temperatures ($t_{h,in}$) increased, the gradient of the fouling resistance increased due to higher surface temperature. Since $t_{s,in}$ has a large effect on the interface temperature between the fouling layer and the solution, it has a greater effect than the $t_{h,in}$ on value of asymptotic fouling resistance.

In the following, the temperature effects on the $CaCO_3$ crystallization fouling under different heat transfer conditions is investigated. Najibi et al. [39] stated the surface temperature effects on the formation of $CaCO_3$ fouling under subcooled flow boiling conditions. The bulk temperature in this experiment was assumed to

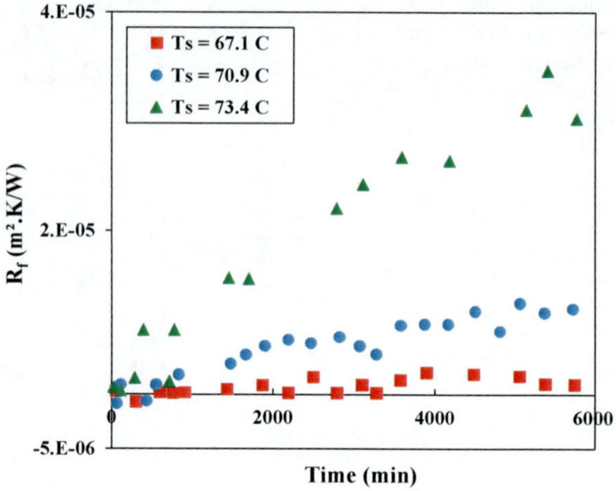

Fig. 3.21 Surface temperature effect on $CaSO_4$ fouling resistance (adapted from [35])

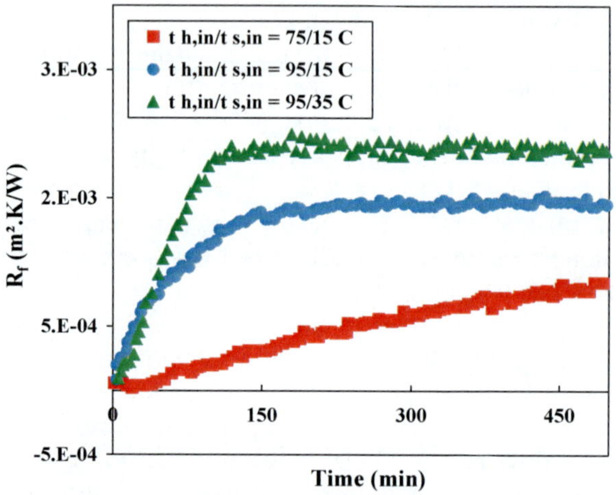

Fig. 3.22 Hot and cold inlet fluid temperature effects on $CaSO_4$ fouling resistance (adapted from [30])

be constant at $T_b = 80$ °C. Other operating conditions of this study can be seen in Table 3.1. Figure 3.23 shows the results of this study. They related the temperature effects and fouling control mechanism to the flow velocity. They stated that the effect of surface temperature in these conditions, unlike $CaSO_4$ salt, was negligible, but with flow velocity increasing, the surface temperature effect was more significant due to the mechanism change to reaction-controlled fouling. In addition, at fouling controlled by mass transfer, increasing surface temperature decreased solubility of salt and increased the mass transfer coefficients, thus, increased the deposition rate, but this increase was much less compared to that observed at higher velocities.

Xiaokai et al. [42] investigated the effect of bulk temperature on $CaCO_3$ fouling formation under subcooled flow boiling conditions. This study was performed at $T_s = 111.2$ °C, $u = 0.9$ m/s, and concentration $[CaCl_2] = 10$ mmol/l and $[NaHCO_3]$

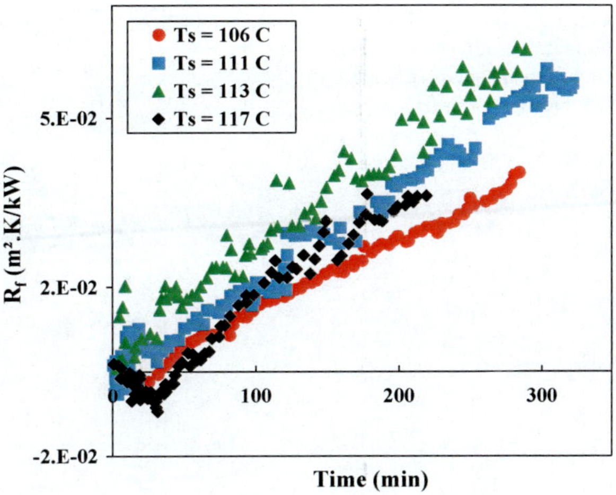

Fig. 3.23 Surface temperature effect on $CaCO_3$ fouling resistance (adapted from [39])

= 20 mmol/l. Figure 3.24 shows the results of this research which was done at two concentrations; the lower concentration is not shown here. In this study, unlike Najibi et al. [39], the fouling control mechanism was determined based on the salt concentration. By comparing the results at different concentrations, they showed that in the lower concentration experiment, the fouling process is controlled by surface reaction that was almost independent of bulk temperature. However, in higher concentration experiments, fouling process is controlled by mass transfer and fouling resistance increased with increasing bulk temperature.

Teng et al. [50] researched inlet hot fluid temperature variation effect on $CaCO_3$ fouling in a double-tube heat exchanger at 25 °C inlet solution temperature. Other information is available in Table 3.1. The results of this research are presented in Fig. 3.25. This study was performed under forced convection conditions. $CaCO_3$ is

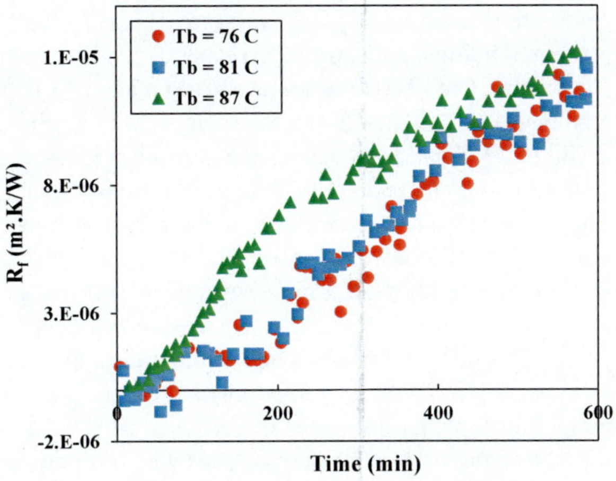

Fig. 3.24 Bulk temperature effect on $CaCO_3$ fouling resistance (adapted from [42])

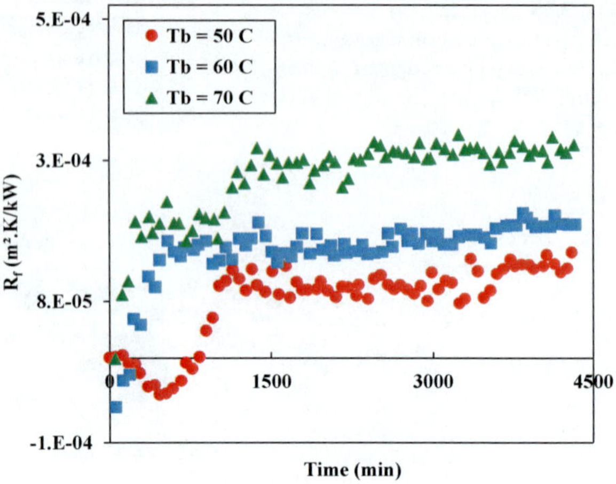

Fig. 3.25 Inlet hot fluid temperature effect on $CaCO_3$ fouling resistance (adapted from [50])

reversing soluble with temperature. Therefore, it can be said that temperature changes have a significant effect on the amount of $CaCO_3$ deposits [40, 77]. At a temperature of 50 °C, an induction period and then a rapid increase in fouling resistance were observed, and finally the fouling resistance reached an asymptotic state. At 70 °C there was almost no sign of an induction period, and the fouling resistance started to increase rapidly at the beginning of the test. The curve reached an asymptotic state at relatively similar times, but the asymptotic value fouling resistance increased by increasing temperature.

Zhenhua et al. [43] researched the wall temperature effects on fouling of $CaCO_3$ during forced convection heat transfer condition. As shown in Fig. 3.26, with an increase of 20 °C in the wall temperature (44–64 °C), the fouling resistance almost doubled. With increasing the wall temperature due to its effect on $CaCO_3$ supersaturation, the fouling resistance increases [40, 77].

Helalizadeh et al. [52] studied surface temperature and bulk temperature effects on crystallization fouling of a mixture of $CaCO_3$ and $CaSO_4$ salts. Their results are presented in Figs. 3.27 and 3.28, respectively. In this study, the salt concentration was constant at $[CaCO_3] = 0.7$ g/l and $[CaSO_4] = 1.5$ g/l. The velocities in Figs. 3.27 and 3.28 were constant at 0.5–0.6 m/s, respectively. In Fig. 3.27 the $T_b = 80$ °C. As shown in Fig. 3.27, the fouling resistance depended strongly on the heat transfer T_s. Supersaturation of salts and consequently fouling resistance increased with increasing T_s. They stated that for velocities below 0.8 m/s (because the fouling controls by mass transfer), the increasing the surface temperature increases fouling resistance. For flow velocities above 0.8 m/s (because approaching the Arrhenius relationship for the fouling controls by reaction), surface temperature effect become increasingly powerful. Figure 3.28 shows bulk temperature effect on fouling resistance. Contrary to expectation, fouling in subcooled flow boiling conditions is dependent to bulk temperature, which indicates the absence of fully developed conditions. Therefore, increasing bulk temperature increases wall temperature and nucleation sites number and finally fouling amount.

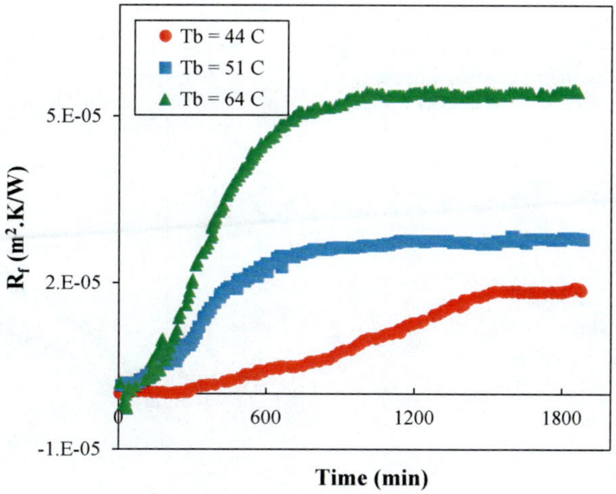

Fig. 3.26 Surface temperature effect on $CaCO_3$ fouling resistance (adapted from [43])

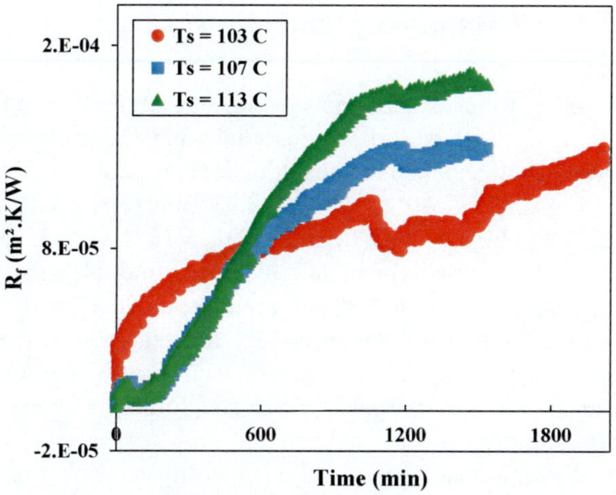

Fig. 3.27 Surface temperature effect on mixed salts fouling resistance (adapted from [52])

In general, it can be said that since $CaSO_4$ and $CaCO_3$ have inverse solubility characteristics with temperature, the crystallization fouling of these salts is strongly affected by temperature. Increases of the reaction rate constant and mass transfer coefficient with increasing temperature are exponential and linear, respectively. On the other hand, according to the reviewed articles, increasing surface temperature increases the number of nucleation sites. Hence, the fouling rates are strongly dependent on the surface temperature especially for the experiments at higher velocities (which are related to the reaction-controlled fouling process). In addition, the dependency of the crystallization fouling on the bulk temperature relative to the surface temperature is negligible. Another point is the increase in the fouling resistance of salts during subcooled flow boiling compared to forced convection due to the presence of bubbles.

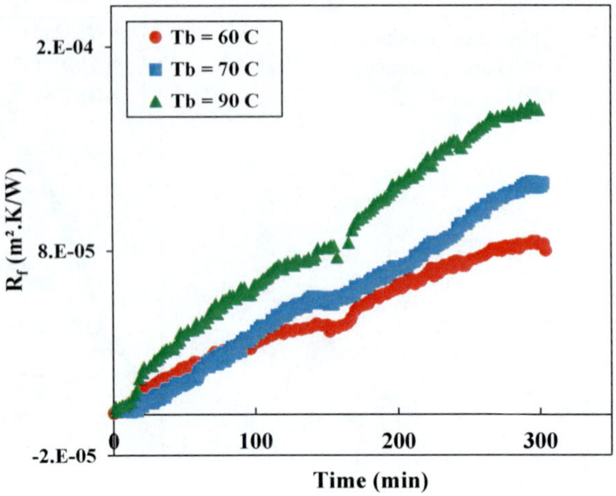

Fig. 3.28 Bulk temperature effect on mixed salts fouling resistance (adapted from [52])

3.4 Effect of Concentration

In order to investigate the salt concentration effect on the fouling resistance during fouling growth period, various studies have been classified and reviewed based on the type of salt and then the heat transfer conditions. Results of this review are presented in Figs. 3.29, 3.30, 3.31, 3.32, 3.33, 3.34, 3.35, 3.36, 3.37, 3.38, 3.39 and 3.40. As shown in Figs. 3.29, 3.30, 3.31, 3.32, 3.33, 3.34, 3.35, 3.36, 3.37, 3.38, 3.39 and 3.40, regardless of the type of salt, increasing the salt concentration enhanced the fouling resistance. In the case of salt mixtures, the increasing of $CaCO_3$ concentration effect on the fouling resistance is very noticeable, and the fouling resistance increases steadily with increasing the concentration of $CaCO_3$. For $CaSO_4$ salt, studies in subcooled flow boiling [7, 57] and forced convection [8, 35, 65, 70] have been investigated.

Peyghambarzadeh et al. [57] investigated effect of three different concentrations of $CaSO_4$ on crystallization fouling under subcooled flow boiling conditions. Results of this study are presented in Fig. 3.29. They have showed that reason for increase in fouling resistance with increasing bulk concentration is the increase in chemical potential difference between the substance in crystal and solution. The formation of crystals larger than the critical radius that are not redissolved in solution and development of crystal lattice as well as increase in supersaturation are all caused by increases in concentration. However, at high concentrations, fouling resistance may be reduced due to bulk precipitation and the formation particulate fouling which are usually loose deposits. In general, supersaturation is the main driving force in the crystallization fouling process, therefore, with increasing the salt concentration, initial nucleation of salts, supersaturation and consequently, the amount of deposition increase [28, 65, 77].

In another study, Najibi et al. [7] studied the $CaSO_4$ concentration effect at two different velocities under subcooled flow boiling conditions. The results of this study

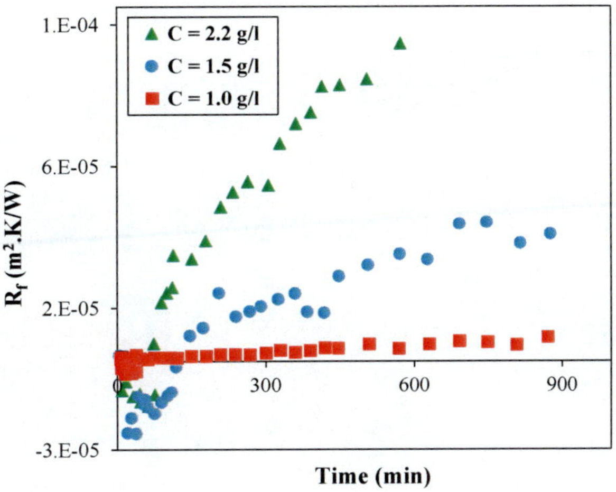

Fig. 3.29 $CaSO_4$ concentration effect on fouling resistance (adapted from [57])

are presented in Figs. 3.30 and 3.31. Table 3.1 are shown the experiment conditions of this research. The velocities of flow are u = 1.2 m/s and u = 0.6 m/s in Figs. 3.30 and 3.31, respectively. Graphs have shown that increasing the salt concentration has a significant effect on the fouling resistance and increased it rapidly. They stated that the fouling rate usually follows Eq. (2.3) [37]. Hasson et al. [37] based on Eq. (2.3) stated that regardless of the fouling mechanism, the effect of concentration is very strong. In addition, at both velocities studied in this work, the reaction mechanism dominated and the fouling resistance increased when the salt concentration increased.

Concentration effect on fouling resistance under forced convection conditions is discussed below. Han et al. [33] studied the $CaSO_4$ concentration effects on fouling in a rectangular channel with vortex producers. The operating conditions of the research are presented in Table 3.1. Figure 3.32 presents the results of this research.

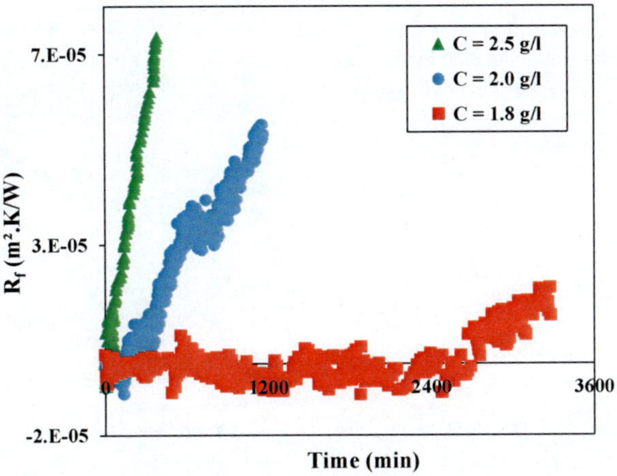

Fig. 3.30 $CaSO_4$ concentration effect on fouling resistance at u = 1.2 m/s (adapted from [7])

Fig. 3.31 $CaSO_4$ concentration effect on fouling resistance at u = 0.6 m/s (adapted from [7])

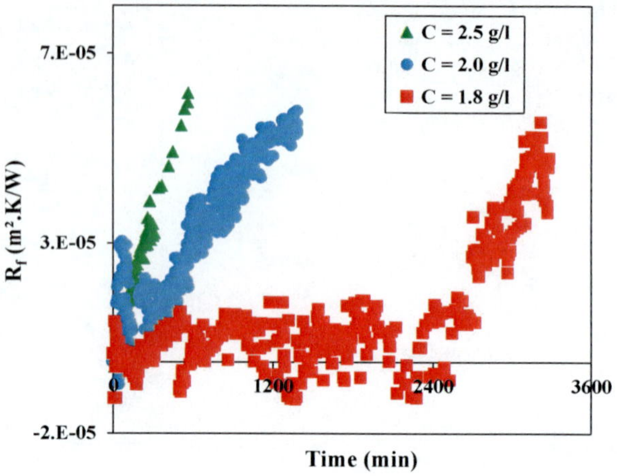

As mentioned earlier, fouling resistance is the result of two simultaneous processes (deposition and removal), before fouling process reaches equilibrium stage of these two processes (asymptotic period), the fouling resistance gradient is directly related to the increase in salt concentration, because increasing the number of foulant ions increases the reaction rate and consequently increases the deposition rate. In addition, with increasing $CaSO_4$ concentration in bulk flow region, the difference between the bulk region and the boundary layer according to Eq. (2.3) becomes large, and this increases the fouling rate [11, 30, 78].

In another study on forced convection conditions, Müller-Steinhagen [8] investigated $CaSO_4$ concentration effect on crystallization of cooling water in heat exchangers. Table 3.1 and Fig. 3.33 are shown the test operating conditions and research results respectively. He stated that since the $CaSO_4$ crystallization is controls

Fig. 3.32 $CaSO_4$ concentration effect on fouling resistance in a rectangular channel with vortex producers (adapted from [33])

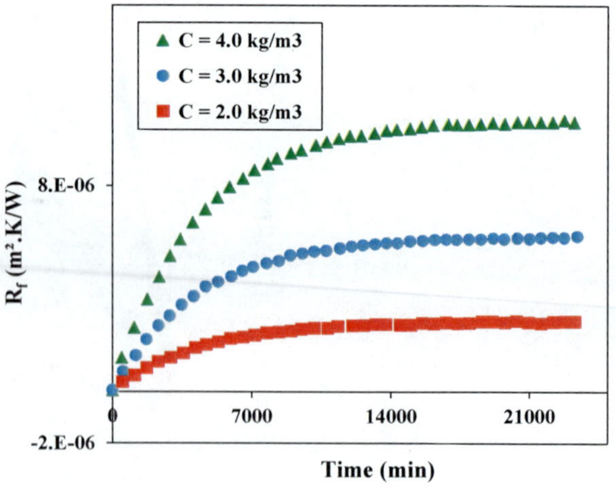

Fig. 3.33 $CaSO_4$ concentration effect on crystallization fouling resistance of cooling water in heat exchangers (adapted from [8])

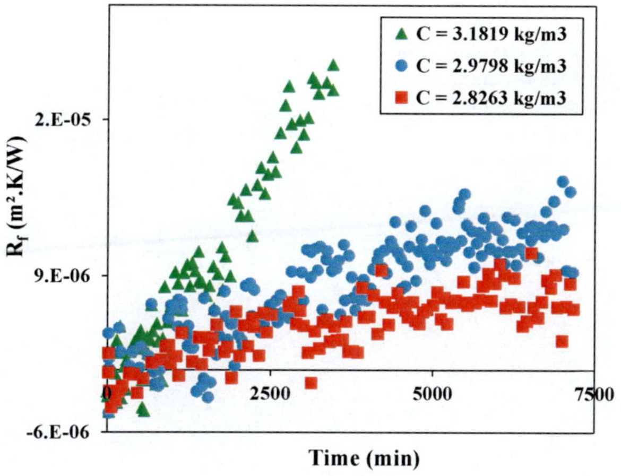

by reaction, increasing concentration increases the fouling resistance. He has shown that after some time of the fouling process, the salt concentration is no longer as important as in the initial stages of the fouling process, and other factors such as shear stress and interface temperature are important due to the reduction of the channel cross section by fouling. In general, with increasing the concentration of foulant in bulk flow the fouling increased linearly.

In another study, Bansal and Müller-Steinhagen [70] in plate heat exchanger studied concentration changes effect on $CaSO_4$ crystallization fouling. The results are shown in Fig. 3.34. The study was performed at two initial concentrations of 2988 and 3198 ppm under forced convection conditions. Since the fouling process of $CaSO_4$ is controlled by a second-order reaction, the driving force of the surface reaction increases with increasing salt concentration, and as a result, the fouling rate increases. Similar to Müller-Steinhagen's [8] research, the effect of concentration at the beginning of the fouling tests is greater, and fouling resistance is no longer affected by salt concentration after reaching an almost constant value.

Maddahi et al. [35] investigated $CaSO_4$ concentration changes effect on the fouling resistance along a solid–liquid fluidized bed heat exchanger with cylindrical particles. Bed voidage and average wall temperature were fixed at 0.65 and 70.8 °C, respectively. According to Fig. 3.35, increasing the salt concentration from 1.2 to 1.9 kg/m^3 increased the fouling resistance significantly. At the concentration of 1.2 kg/m^3, the fouling resistance fluctuated around zero, so, the concentration of the foulant fluid at the wall temperature was not supersaturated. Supersaturation conditions were provided with increasing the salt concentration.

Studies on salt concentration effect on fouling resistance under pool boiling conditions are reviewed here. Malayeri et al. [14] and Malayeri and Müller-Steinhagen [17] investigated $CaSO_4$ deposition under pool boiling conditions. They showed that when the salt concentration increased from 0.25 to 1.5 g/l, the nucleation sites density decreased and the bubbles diameter increased. On the other hand, increasing the salt concentration causes a change in the fluid physical properties such as surface

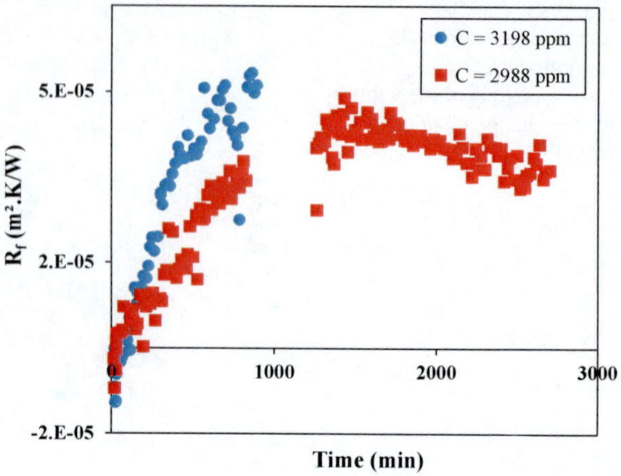

Fig. 3.34 $CaSO_4$ concentration effect on fouling resistance in plate heat exchanger (adapted from [70])

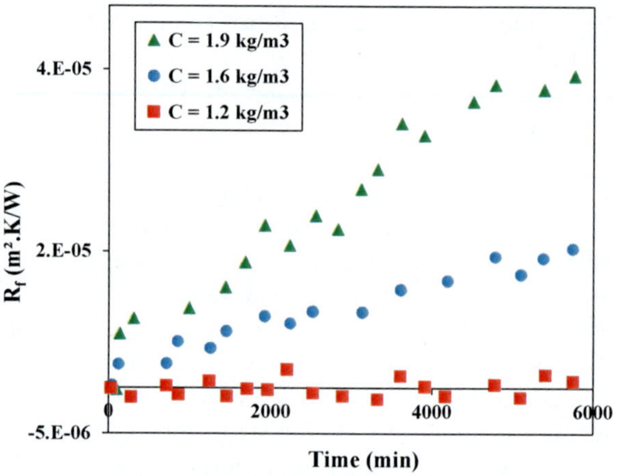

Fig. 3.35 $CaSO_4$ concentration effect on fouling resistance along a solid–liquid fluidized bed heat exchanger with cylindrical particles (adapted from [35])

tension and as a result, the formation mechanism of bubble changes and these factors reduce the heat transfer coefficient. As the $CaSO_4$ concentration increases, the heat transfer coefficient due to the supersaturation of the salt solution at the triple interface decreases rapidly. Sarafraz et al. [79–86] and Peyghambarzadeh et al. [87] later presented similar results under pool boiling conditions.

In order to investigate the $CaCO_3$ concentration effects on the crystallization fouling, studies on the different heat transfer conditions are reviewed. The results of Najibi et al. [39] in subcooled flow boiling, Teng et al. [50] in forced convection, and Cai et al. [48] in pool boiling for $CaCO_3$ salt solution are presented here.

Figure 3.36 related to the results of Najibi et al. [39] shows that concentration plays a decisive role in the deposition process. The conditions of their tests are presented in Table 3.1. They said that bulk concentration could affect the amount of deposit as well as the initial delay time. They reported that supersaturation is the major reason of the fouling. They examined the concentration effect at different velocities, the results for $CaCO_3$ salt have a similar trend to those for $CaSO_4$ salt, and they concluded that the concentration effect became more pronounced at higher velocities, indicating a change in the fouling mechanism from diffusion to the reaction.

Teng et al. [50] studied the $CaCO_3$ concentration effect on the fouling in a double-pipe heat exchanger (forced convection). The results of their research are presented in Fig. 3.37. Table 3.1 presents the test performance conditions. Results showed that increasing the concentration of solution increased the fouling resistance and decreased the heat transfer rate. It indicates faster formation of the fouling layer at higher concentrations rather than at lower concentrations. Figure 3.37 show that by increasing the salt concentration (by 300–500 mg/l), the fouling resistance almost doubled. By increasing the salt concentration, supersaturation is achieved faster and the circulation of the solution in the apparatus system and heat exchanger increases the rate of nucleation and causes $CaCO_3$ deposition.

The results of Scanning Electron Microscope (SEM) proved that the calcite crystals with orthorhombic crystal structure are present at concentrations of 300 and 400 mg/l, which leads to a stronger structure. Kazi et al. [88] previously observed similar results.

Cai et al. [48] investigated $CaCO_3$ concentration effect on fouling in pool boiling condition. Figure 3.38 shows the results of this research. The initial concentration of $CaCO_3$ significantly affected the fouling resistance, and the higher concentration of $CaCO_3$ resulted in a higher asymptotic fouling resistance. During the experiment, as the concentration of the precipitating salt decreased, the increasing rate of the fouling resistance decreased. It should be mentioned that due to the low solubility of $CaCO_3$, its crystals settled on the bottom of the test pool.

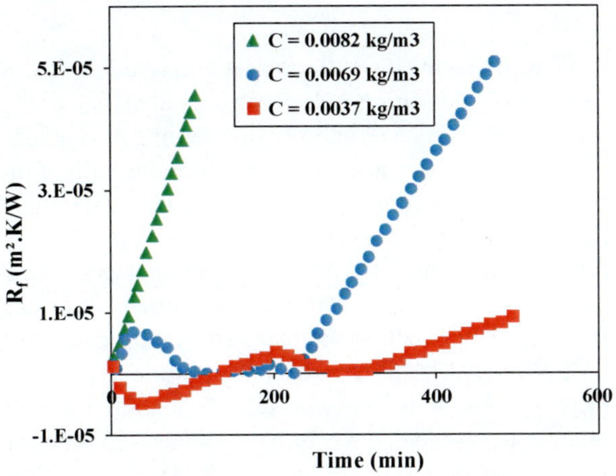

Fig. 3.36 $CaCO_3$ concentration effect on fouling resistance (adapted from [39])

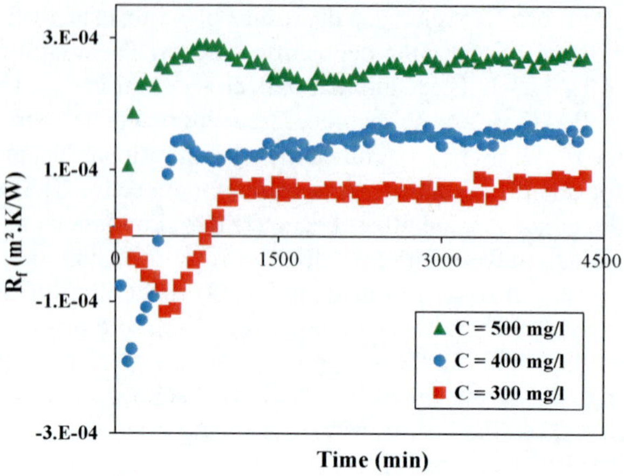

Fig. 3.37 $CaCO_3$ concentration effect on fouling resistance in a double-pipe heat exchanger (adapted from [50])

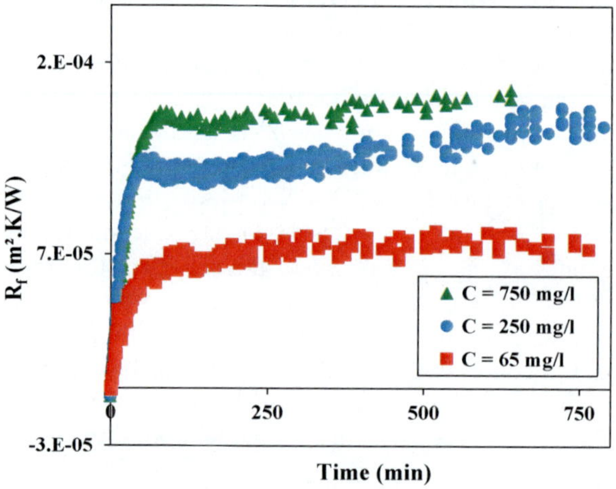

Fig. 3.38 $CaCO_3$ concentration effect on fouling resistance (adapted from [48])

Comparing the three aforementioned studies on $CaCO_3$ deposition under different heat transfer conditions, it can be concluded that in subcooled flow boiling conditions, due to the presence of bubbles, the fouling resistance increases linearly. However, in the other two cases after the rapid initial growth of the fouling resistance, especially in pool boiling conditions where the curves are initially sharply increased, the curves become asymptotic even at the highest concentration studied. In subcooled flow boiling conditions, by increasing the salt concentration, the slope of the fouling curve is much higher and unlike two other heat transfer modes, it does not tend to be asymptotic at all, in fact, the removal process does not occur in this condition.

To study the mixed salt concentration effect on the fouling resistance, results of two different studies, one under subcooled flow boiling conditions [52] and the other under forced convection conditions [56], are presented. Helalizadeh et al. [52] studied concentration of $CaCO_3$ and $CaSO_4$ salts effect at $u = 0.5$ m/s, $T_b = 80$ °C and q = 300 kW/m^2 under subcooled flow boiling conditions. The results are presented in Fig. 3.39. Fouling resistance curves did not follow a specific trend. The lowest fouling resistance was related to pure $CaSO_4$ deposition, and increasing the concentration of $CaCO_3$ in the mixed salt solution increased the fouling resistance. The highest fouling resistance was related to the highest salt concentrations ($CaSO_4$: 2 kg/m^3, $CaCO_3$: 1 kg/m^3). It can be said that increasing the concentration of $CaCO_3$ greatly increased the fouling resistance and changed the shape of the fouling curve from quasi-asymptotic to linear with a high slope.

Song et al. [56] studied $CaCO_3$ and $CaSO_4$ deposition in plate heat exchanger at different salt concentrations. The heat transfer conditions were forced convection. Their results (Fig. 3.40) show that with increasing fouling, the removal rate also increases and the fouling curve becomes asymptotic, they have stated that deposit particles can cause particulate fouling and act as nucleation sites thus increase the deposition rate [89]. In addition, it can be seen that due to the higher adhesion of $CaCO_3$ than $CaSO_4$, its effect on fouling resistance is greater. Also, the fouling resistance of the salt mixture is higher than the fouling resistance of the pure salts, which may be due to the higher nucleation sites of the salt mixture.

In all reviewed studies, regardless of the type of salt, increasing the salt concentration increased the fouling resistance due to the increase in supersaturation. Therefore, the driving force of the fouling increases and more crystals grow on the surface. In the mixed salts crystallization fouling, it is clear that increasing $CaCO_3$ concentration, increased the fouling resistance much faster. In this case, due to the stronger crystalline structure of $CaCO_3$, increasing the concentration of this salt changed the shape of the fouling resistance curve from asymptotic shape to linear trend.

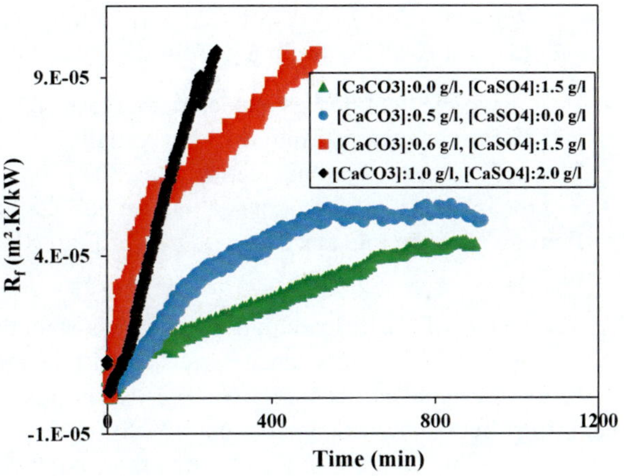

Fig. 3.39 Mixed salts concentration effect on fouling resistance (adapted from [52])

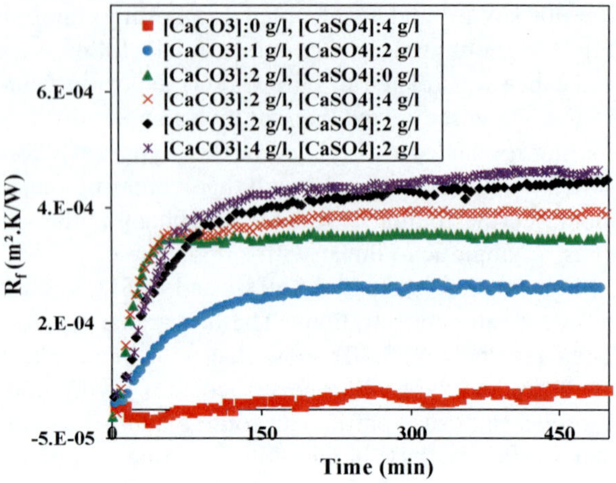

Fig. 3.40 Mixed salts concentration effect on fouling resistance (adapted from [56])

3.5 Effect of Heat Flux

As mentioned earlier, few studies have been performed on the effect of heat flux on the crystallization fouling resistance. A number of these studies are presented in Figs. 3.41, 3.42, 3.43, 3.44 and 3.45. The reviewed studies are presented for $CaSO_4$ salt only in pool boiling conditions [24, 90, 91] and for $CaCO_3$ salt in pool boiling [48] and forced convection heat transfer [44, 47].

Esawy [90] investigated the effect of heat flux on the $CaSO_4$ fouling resistance in stainless steel tubes. He showed that the T_s for heat fluxes of 200 kW/m^2 and 300 kW/m^2 exceeded 170 °C after 24 and 15 h, respectively. He also described the effect of heat flux on the fouling as follows:

- Increasing the heat flux as expected, increases the T_s, this increase in T_s leads to an increase in supersaturation and eventually an increase in fouling rate.
- The activation of more nucleation sites, the production of more vapour bubbles, and the increase in deposit beneath these bubbles are all due to the increase in T_s.
- Fouling reaction rate at surface (k_r) according to Eq. (2.1) is strongly affected by the T_s.

The results of this research on the stainless-steel tube are presented in Fig. 3.41. As shown, the fouling resistance increased with increasing heat flux.

In another study under pool boiling conditions, Esawy et al. [24] studied the sintering effect on fouling formation of $CaSO_4$ solutions. Results of this research are presented in Fig. 3.42. Figure 3.42 shows that, as expected, a lower fouling resistance is obtained due to the presence of fewer vapors bubbles in the lower heat flux. They express the behaviors of fouling curves influenced by T_s fouling layer. They have shown that when the fouling layer temperature is higher than the minimum sintering

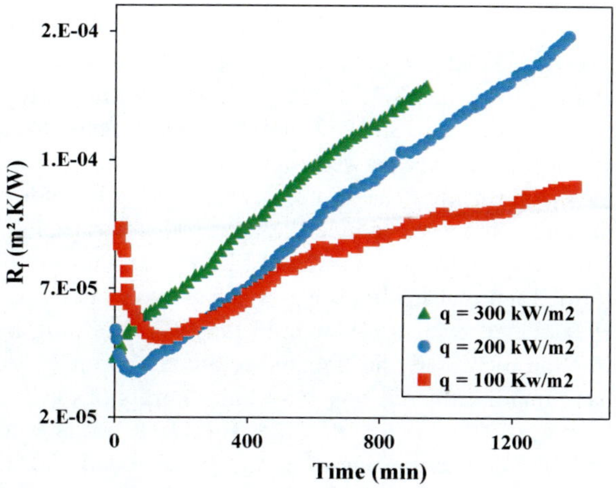

Fig. 3.41 Heat flux effect on $CaSO_4$ fouling resistance on stainless-steel tube (adapted from [90])

temperature (MST) (under nuclear boiling conditions), the fouling curve becomes asymptotic and if the fouling layer temperature is lower than MST, the fouling curve will continue to grow without asymptote.

In addition, they showed that heat flux increasing leads to a reduction in asymptotic value, a result that was previously reported by Jamialahmadi and Müller-Steinhagen [13]. On the other hand, they stated that due to the presence of salt, the quantity of nucleation sites decreased but the outgoing bubbles diameter increased. As the heat flux increases for both distilled water and $CaSO_4$ solution, the bubble nucleation sites density increases. Heat flux and foulant concentration strongly affected the heat transfer coefficient.

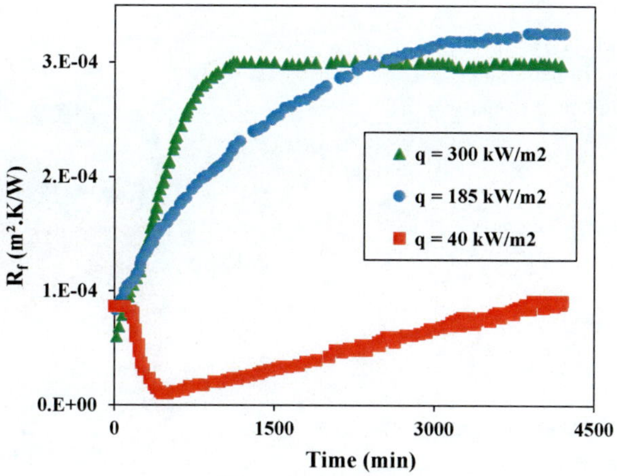

Fig. 3.42 Heat flux effect on $CaSO_4$ fouling resistance (adapted from [24])

In another study, the Jamialahmadi and Müller-Steinhagen [91] results (at several heat fluxes and pool boiling condition) showed that the fouling layer became denser at higher heat fluxes. In addition, bubbles formed at the interface between the fouling layer/solution instead of the metal surface. Because smaller bubbles are formed due to the large difference between the surface tension of the solution and the salt deposit and the metal surface. The homogeneous fouling layer development and the inactivation of nucleation sites reduce the heat transfer rate. For subcooled flow boiling heat transfer conditions, Bansal and Müller-Steinhagen [70] stated that fouling removal formed at high heat fluxes was much more difficult than the fouling at low heat fluxes. Helalizadeh et al. [54] implied that by increasing heat flux, the nucleate boiling fraction increased and fine and relatively compact fouling appeared. The heat flux has a major influence on the fouling stability of $CaCO_3$ solutions and $CaCO_3/CaSO_4$ mixtures. They also stated that $CaSO_4$ deposits were broken at high heat fluxes and high concentrations when the T_s exceeded 150 °C probably due to the bubbles generation under the fouling layer. In this condition, the shear stress introduced by the bubbles rising and causes the saw-tooth curve was formed. Vosough et al. [4, 36] also showed this phenomenon.

For $CaCO_3$ salt under boiling pool conditions, the results of the study by Cai et al. [48] are presented in Fig. 3.43. Test operating conditions are shown in Table 3.1. As shown in Fig. 3.43, the fouling resistance curves at different heat fluxes had a similar trend. Increasing the heat flux increased the fouling resistance.

Pääkkönen et al. [44] studied the crystallization fouling of $CaCO_3$ under forced convection conditions. The results of this study are shown in Fig. 3.44. They showed that heat flux is directly related to Ts and Ts is directly related to the reaction rate constants and supersaturation (that are the main cause of fouling growth), so heat flux will also directly affect fouling resistance, as a result, increasing the heat flux increases the rate of the linear growth of fouling resistance.

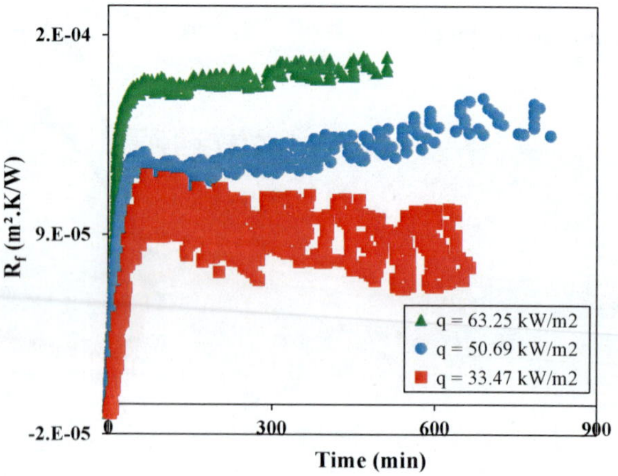

Fig. 3.43 Heat flux effect on $CaCO_3$ fouling resistance (adapted from [48])

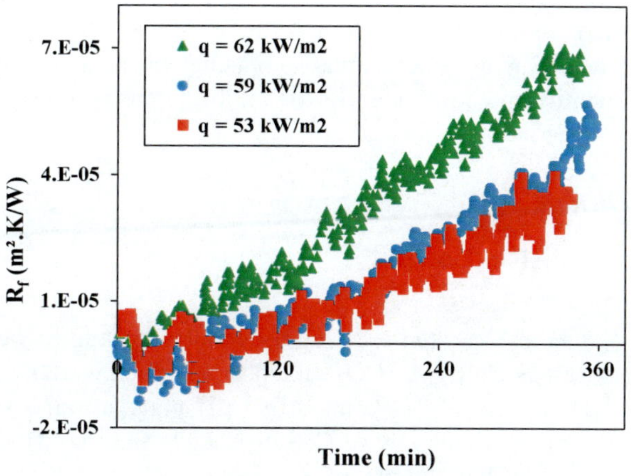

Fig. 3.44 Heat flux effect on $CaCO_3$ fouling resistance (adapted from [44])

Al-Hadhrami et al. [47] investigated the $CaCO_3$ fouling in a twisted tube heat exchanger. The condition of this experiment was also forced convection. They investigated the effect of different heat inputs of 200–800 W at different operating conditions (Fig. 3.45). Other operating conditions are listed in Table 3.1. According to Fig. 3.45, fouling resistance is very sensitive to changes in the heat input and is steadily increasing as the inlet heat increases. However, a noteworthy point in this study (which is not presented in this diagram) is the lack of significant effect of incoming heat fluxes (mainly at high inlet heat) on fouling resistance at high flow velocities.

In general, it can be concluded that the result of increasing the heat flux for both salts was an increase in the fouling resistance. The reason for this phenomenon is the increase in Ts as a result of the decrease in salt solubility and the consequent

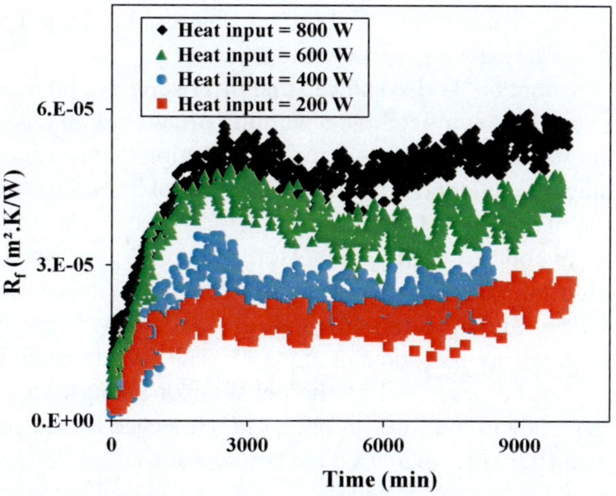

Fig. 3.45 Heat input effect on $CaCO_3$ fouling resistance (adapted from [47])

increase in supersaturation at the interface, all of which increase the fouling resistance. On the other hand, increasing the heat flux increases the nucleation sites quantity and consequently, the bubbles density, the result of this phenomenon is the fouling intensification.

3.6 Effect of pH

In the previous studies, a limited number of studies were performed on $CaCO_3$ salts as well as $CaSO_4/CaCO_3$ mixed salts, which were reviewed below.

Höfling et al. [92] stated that pH plays a major role in determining the deposition rate, especially in $CaSO_4/CaCO_3$ fouling. The minimum fouling appears at pH = 7. When the pH < 7 (acidic) or pH > 7 (alkaline), the fouling conditions are quite different. At pH less than 6, just $CaSO_4$ is visible by X-Ray Diffraction (XRD). For higher pH values, both $CaCO_3$ and $CaSO_4$ salts precipitated in different percentages. In upper and middle layers at pH = 7 and then pH = 6, the highest resistance of crystals is formed. However, precipitation was less likely to occur at these two pH values, terms of molecular bonding, the occurrence of crystallization was very in likely. Augustin and Bohnet [73] also studied the pH effect on $CaCO_3$ fouling. They express the saturation index in three ways: (1) the number of free hydrogen ions divided by the calcium ions, (2) the amount of activity product divided by the solubility product and (3) the actual pH subtraction of saturation pH.

One aspect of the deposition problem is the effect of pH on the crystallization. Simultaneously, for pH > 6, the amount of $CaCO_3$ precipitation increases. Supersaturation is the driving force of crystallization. Figure 3.46 indicates that the asymptotic fouling resistance with increasing pH from 6.2 to 7.1 at a constant u = 0.5 m/s increased. When the pH > 6, the supersaturation increases due to the increase in the number of free carbonate ions, resulting in the asymptotic amount of fouling resistance increases. For $CaCO_3$ salt, increasing the pH value increased the saturation index and hence, increased the fouling.

Bohnet [93] also showed that increasing pH, increased $CaCO_3$ asymptotic fouling resistance owing to higher stability of the fouling layer owing to the more nucleation sites at higher supersaturation. Saturation theory states that the saturation of $CaCO_3$ single compound increases at a higher pH. $CaSO_4$ saturation was almost constant in the pH range tested.

In contrast to this study, Höfling et al. [92] showed that at a certain concentration of salt mixture, by increasing pH from 4 to 7, the fouling resistance decreased continuously, and the lowest fouling resistance was observed at pH = 7. The results of this research are presented in Fig. 3.47. The fouling behavior of $CaSO_4/CaCO_3$ solutions cannot be described by the behavior of a single salt. By increasing pH value, less tendency to fouling was observed. However, increasing $CaCO_3$ salt concentration at a constant pH, increased the amount of fouling. In addition, Bramson et al. [94] and Wang et al. [49] stated that $CaCO_3$ increased the strength of $CaSO_4$ fouling layers. Crystalline transformation at pH = 4–5.5 drastically increases the strength of fouling

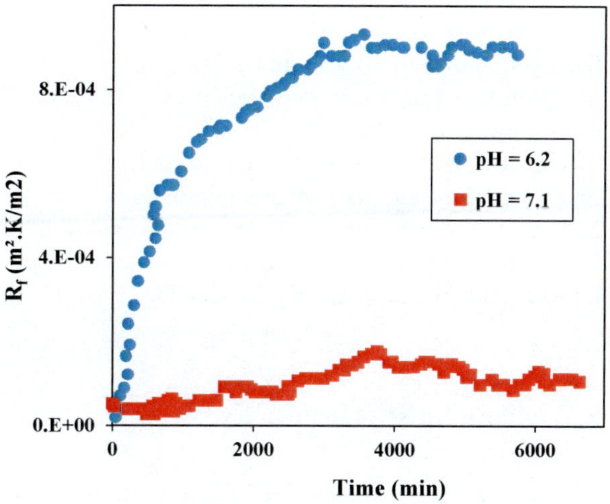

Fig. 3.46 pH effect on $CaCO_3$ fouling resistance (adapted from [73])

layers. In addition, Sudmalis and Sheikholeslami [55] stated that in some cases the equilibrium concentration of pure $CaCO_3$ was higher than that of pure $CaSO_4$. This is because the solubility of $CaCO_3$ is dependent on pH while the solubility of $CaSO_4$ (at $pH > 3$) is pH independent; therefore, the co-existence effect of anions that precipitate calcium cannot be ignored. Höfling et al. [92] observed the effect of pH not only on the fouling resistance but also on the fouling layer structure. By increasing pH values, less fouling tendency was observed. Alahmad [20] implied that $CaSO_4$ fouling at $pH > 7$ have always been present. On the other hand, corrosion increased rapidly at low pHs. Therefore, he has suggested precise pH control between 6 and 7 to prevent fouling and corrosion problems.

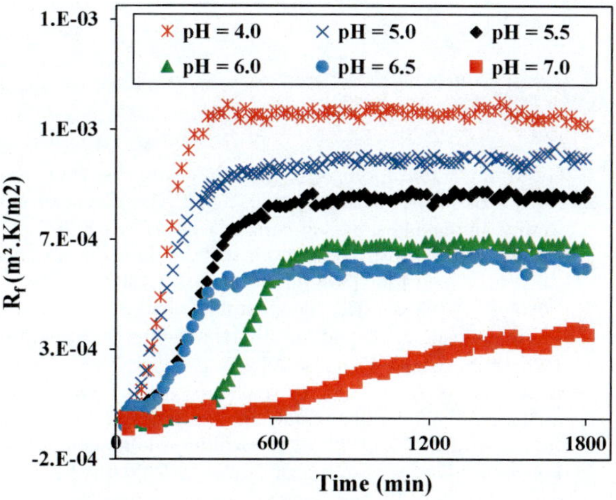

Fig. 3.47 pH effect on mixed fouling resistance (adapted from [92])

In most of the reviewed studies on the effect pH on the fouling resistance, it has been stated that by increasing the pH to 7, the amount of fouling, especially in the mixed salt, decreased. Since the solubility of $CaCO_3$ is pH-dependent but the solubility of $CaSO_4$ (for pH > 3) is pH-independent, therefore, the behavior of mixed salt at different pH values were affected by the solubility of $CaCO_3$ and the co-existence of anions that precipitated calcium. As the pH increases, the amount of $CaCO_3$ fouling increases due to the increase in the saturation index. Therefore, increasing the pH value decreases the amount of mixed salt crystallization fouling but increases the amount of pure $CaCO_3$ crystallization fouling.

3.7 Concluding Remarks

Despite numerous studies, crystallization fouling on the heat transfer surface is still a complex phenomenon. The effect of different operating parameters on the crystallization fouling also requires more comprehensive studies. In the present study, a complete review of the previous studies has been performed with emphasis on the crystallization fouling on the heat transfer surfaces. Then, the effect of various operational parameters including flow velocity, bulk temperature, surface temperature, fluid bulk concentration, heat flux and pH value on the crystallization fouling in the last decades have been reviewed. In the present study, the lack of comprehensive studies on the effect of various parameters such as flow velocity, surface temperature and pH on the different salts, especially the two salts in question ($CaSO_4$ and $CaCO_3$) have been clearly identified. On the other hand, studies on the mixed salts fouling ($CaSO_4/CaCO_3$) and the impacts of operational parameters on this type of fouling have been very rare. In addition, the effect of pH on the crystallization fouling of different salts needs further investigation.

References

1. Gilmour CH. No fooling-No fouling. Chem Eng Prog 1965; 61(7): 49–54.
2. Bansal B, Müller-Steinhagen H, Chen XD. Effect of suspended particles on crystallization fouling in plate heat exchangers. J Heat Trans 1997; 119(3): 568–574.
3. Geddert T, Augustin W, Scholl S. Induction time in crystallization fouling on heat transfer surfaces. Chem Eng Technol 2011; 34(8): 1303–1310.
4. Vosough A, Assari MR, Peyghambarzadeh SM, Azizi S. Influence of fluid flow rate on the fouling resistance of calcium sulfate aqueous solution in subcooled flow boiling condition. Int J Therm Sci 2020; 154: 106397.
5. Hasson, D., Zahavi, J., 1970, Mechanism of Calcium Sulfate Scale Deposition on Heat Transfer Surfaces, I&EC Fundamentals, 9(1), pp. 1–10.
6. Jamialahmadi, M., Blochl, R., Müller-Steinhagen, H., 1989, Bubble Dynamics and Scale Formation During Boiling of Aqueous Calcium Sulphate Solutions, Chemical Engineering and Processing: Process Intensification, 26, pp. 15–26.
7. Najibi, S.H., Müller-Steinhagen, H., Jamialahmadi, M., 1997, Calcium sulphate scale formation during subcooled flow boiling, Chemical Engineering Science, 52(8), 1265–1284.

8. Müller-Steinhagen, H., 1999, Cooling-Water Fouling in Heat Exchangers, Advances in Heat Transfer, 33, 415–496.
9. Förster M, Augustin W, Bohnet M, Influence of the adhesion force crystal/heat exchanger surface on fouling mitigation. Chem Eng Process 1999; 38: 449–461.
10. Müller-Steinhagen H, Zhao Q, Helalizadeh A, Ren XG. The Effect of surface properties on $CaSO_4$ scale formation during convective heat transfer and subcooled flow boiling. Can J Chem Eng 2000; 78: 12–20.
11. Bansal B, Müller-Steinhagen H, Chen XD. Performance of plate heat exchangers during calcium sulphate fouling - investigation with an in-line filter. Chem Eng Process 2000; 39: 507–519.
12. Jamialahmadi M, Helalizadeh A, Müller-Steinhagen H. Pool boiling heat transfer to electrolyte solutions. Int J Heat Mass Tran 2004; 47(4): 729–742.
13. Jamialahmadi M, Müller-Steinhagen H. A new model for the effect of calcium sulfate scale formation on pool boiling heat transfer. J Heat Trans 2004; 126: 507–517.
14. Malayeri MR, Müller-Steinhagen H, Bartlett TH. Fouling of tube bundles under pool boiling conditions. Chem Eng Sci 2005; 60: 1503–1513.
15. Behbahani RM, Müller-Steinhagen H, Jamialahmadi M. Investigation of scale formation in heat exchangers of phosphoric acid evaporator plants. Can J Chem Eng 2006; 84: 189–197.
16. Mwaba MG, Rindt CCM, Van Steenhoven AA, Vorstman MAG. Experimental investigation of $CaSO_4$ crystallization on a flat plate. Heat Transfer Eng 2006; 27(3): 42–54.
17. Malayeri, M. R., Müller-Steinhagen, H., 2007, Initiation of $CaSO_4$ Scale Formation on Heat Transfer Surface under Pool Boiling Conditions, Heat Transfer Engineering, 28(3), 240–247.
18. F. Fahiminia, A.P. Watkinson and N. Epstein, "Early events in the precipitation fouling of calcium sulphate dihydrate under sensible heating conditions," Can. J. Chem. Eng., vol. 85, no. 5, pp. 679–691, 2007.
19. A. Herz, M.R. Malayeri and H. Müller-Steinhagen, "Fouling of roughened stainless steel surfaces during convective heat transfer to aqueous solutions," Energy Conv. Manag.; vol. 49, no. 11, pp. 3381–3386, 2008.
20. M. Alahmad, "Factors affecting scale formation in sea water environments - an experimental approach," Chem. Eng. Technol., vol. 31, no. 1, pp. 149–156, 2008.
21. A. Al-Janabi, M.R. Malayeri and H. Müller-Steinhagen, "Experimental investigation of crystallization fouling on grooved stainless steel surfaces during convective heat transfer," Heat Transfer Eng., vol. 30, no. 10–11, pp. 832–839, 2009.
22. A. Al-Janabi, M.R. Malayeri and O.O. Badran, "Performance of shot peened surfaces subject to crystallization fouling," Int. J. Therm. Sci., vol. 111, pp. 379–389, 2017.
23. Al-Janabi A, Malayeri MR. Innovative non-metal heat transfer surfaces to mitigate crystallization fouling. Int J Therm Sci 2019; 138: 384–392.
24. M. Esawy, M.S. Abd-Elhady, M.R. Malayeri and H. Müller-Steinhagen, "Influence of sintering on deposit formation during pool boiling of calcium sulphate solutions," Exp. Therm. Fluid Sci., vol. 34, no. 8, pp. 1439–1447, 2010.
25. Bansal B, Müller-Steinhagen H, Chen XD. Comparison of crystallization fouling in plate and double-pipe heat exchangers. Heat Transfer Eng 2010; 22: 13–25.
26. Esawy M, Malayeri MR, Muller-Steinhagen H, (2010) Crystallization fouling of finned tubes during pool boiling effect of fin density. Heat Mass Transfer 46: 1167–1176.
27. S.M. Peyghambarzadeh, M.M. Sarafraz, N. Vaeli, E. Ameri, A. Vatani and M. Jamialahmadi, "Forced convective and subcooled flow boiling heat transfer to pure water and n-heptane in an annular heat exchanger," Ann. Nucl. Energy, vol. 53, pp. 401–410, 2013.
28. Peyghambarzadeh, S.M., Vatani, A., Jamialahmadi, M., 2013, Influences of bubble formation on different types of heat exchanger fouling, Applied Thermal Engineering, 50, 848–856.
29. Peyghambarzadeh SM, Bahrami N. Statistical analysis of calcium sulfate scaling under boiling heat transfer. Appl Therm Eng 2013; 53: 108–113.
30. Lee, E., Jeon, J., Kang, H., Kim, Y., 2014, Thermal resistance in corrugated plate heat exchangers under crystallization fouling of calcium sulfate ($CaSO_4$), International Journal of Heat and Mass Transfer, 78, 908–916.

31. M.S. Abd-Elhady, M.R. Jalalirad and M.R. Malayeri, "Influence of injected projectiles on the induction period of crystallization fouling," Heat Transfer Eng., vol. 35, no. 3, pp. 232–245, 2014.
32. Abd-Elhady MS, Malayeri MR. Transition of convective heat transfer to subcooled flow boiling due to crystallization fouling. Appl Therm Eng 2016; 92: 122–129.
33. Z. Han, Z. Xu and J. Wang, "$CaSO_4$ fouling characteristics on the rectangular channel with half-cylinder vortex generators," Appl. Therm. Eng., vol. 128, pp. 1456–1463, 2018.
34. Maddahi MH, Hatamipour MS, Jamialahmadi M. A model for the prediction of thermal resistance of calcium sulfate crystallization fouling in a liquid-solid fluidized bed heat exchanger with cylindrical particles. Int J Therm Sci 2019; 145: 106017.
35. Maddahi MH, Hatamipour MS, Jamialahmadi M. Experimental study of calcium sulfate fouling in a heat exchanger during liquid-solid fluidized bed with cylindrical particles. Int J Therm Sci 2018; 125: 11–22.
36. A. Vosough, S.M. Peyghambarzadeh and M.R. Assari, "Influence of thermal shock on the mitigation of calcium sulfate crystallization fouling under subcooled flow boiling condition," Appl. Therm. Eng., vol. 164, 114434, 2020.
37. D. Hasson, M. Avriel, W. Resnick, T. Rozenman and S. Winderich, "Mechanism of calcium carbonate scale deposition on heat transfer surfaces," Ind. Eng. Chem. Fund., vol. 7, pp. 59–65, 1968.
38. Sheikholeslami R, Watkinson AP. Scaling of plain and externally finned heat exchanger tubes. J Heat Transfer 1986; 108: 147–152.
39. Najibi, S.H., Müller-Steinhagen, H., Jamialahmadi, M., 1997, Calcium Carbonate Scale Formation during Subcooled Flow Boiling, Journal of Heat Transfer, 119, 767–775.
40. Q. Yang, Y. Liu, A. Gu, J. Ding and Z. Shen, "Investigation of induction period and morphology of $CaCO_3$ fouling on heated surface," Chem. Eng. Sci., vol. 57, no. 6, pp. 921–931, 2002.
41. Yang Q, Ding J, Shen Z. Investigation on fouling behaviors of low-energy surface and fouling fractal characteristics. Chem Eng Sci 2000; 55: pp. 797–805.
42. X. Xiaokai, M. Chongfang and C. Yongchang, "Mechanism of calcium carbonate scale deposition under subcooled flow boiling conditions," Chinese J. Chem. Eng., vol. 13, no. 4, pp. 464–470, 2005.
43. Q. Zhenhua, C. Yongchang and M. Chongfang, "Experimental study of fouling on heat transfer surface during forced convective heat transfer," Chinese J. Chem. Eng., vol. 16, no. 4, pp. 535–540, 2008.
44. Pääkkönen, T.M., Riihimäki, M., Simonson, C.J., Muurinen, E., Keiski, R.L., 2012, Crystallization fouling of $CaCO_3$ - Analysis of experimental thermal resistance and its uncertainty, International Journal of Heat and Mass Transfer, 55, 6927–6937.
45. Pääkkönen TM, Riihimäki M, Puhakka E, Muurinen E, Simonson CJ, Keiski RL. Crystallization fouling of $CaCO_3$: effect of bulk precipitation on mass deposition on the heat transfer surface. In: Müller-Steinhagen H, Malayeri MR, Watkinson AP, editors. Proceedings of 8th international conference on heat exchanger fouling and cleaning, Schladming, Austria, 2009, 209–216.
46. L.D. Tijing, H.Y. Kim, D.H. Lee, C.S. Kim and Y.I. Cho, "Physical water treatment using RF electric fields for the mitigation of $CaCO_3$ fouling in cooling water," Int. J. Heat Mass Tran. vol. 53, pp. 1426–1437, 2010.
47. L.M. Al-Hadhrami, A. Ahmad and A. Al-Qahtani, "Experimental study of fouling resistance in twisted tube heat exchanger," Heat Transfer Eng., vol. 33, no. 12, pp. 1024–1032, 2012.
48. Cai, Y., Liu, M., Hui, L., 2013, $CaCO_3$ Fouling on Microscale-Nanoscale Hydrophobic Titania-Fluoroalkylsilane Films in Pool Boiling, AIChE J, 59(7), 2662–2678.
49. L.C. Wang, S.F. Li, L.B. Wang, K. Cui, Q.L. Zhang, H.B. Liu and G. Li, "Relationships between the characteristics of $CaCO_3$ fouling and the flow velocity in smooth tube," Exp. Therm. Fluid Sci., vol. 74, pp. 143–159, 2016.
50. K.H. Teng, S.N. Kazi, A. Amiri, A.F. Habali, M.A. Bakar, B.T. Chewa, A. Al-Shamma, A. Shaw, K.H. Solangi and G. Khan, "Calcium carbonate fouling on double-pipe heat exchanger with different heat exchanging surfaces," Powder Technol., vol. 315, pp. 216–226, 2017.

51. Al-Gailani A, Sanni O, Charpentier TVJ, Crisp R, Bruins JH, Neville A. Inorganic fouling of heat transfer surface from potable water during convective heat transfer. Appl Therm Eng 2020; 184: 116271.
52. Helalizadeh, A., Müller-Steinhagen, H., Jamialahmadi, M., 2000, Mixed salt crystallization fouling, Chem. Eng. Process, 39, 29–43.
53. Helalizadeh A, Müller-Steinhagen H, Jamialahmadi M. Mathematical modeling of mixed salt precipitation during convective heat transfer and subcooled flow boiling. Chem Eng Sci 2005; 60: 5078–5088.
54. Helalizadeh A, Müller-Steinhagen H, Jamialahmadi M. Application of fractal theory for characterization of crystalline deposits. Chem Eng Sci 2006; 61(6): 2069–2078.
55. M. Sudmalis and R. Sheikholeslami, "Coprecipitation of $CaCO_3$ and $CaSO_4$," Can. J. Chem. Eng., vol. 78, no. 1, pp. 21–31, 2009.
56. Song KS, Lim J, Yun S, Kim D, Kim Y. Composite fouling characteristics of $CaCO_3$ and $CaSO_4$ in plate heat exchangers at various operating and geometric conditions. Int J Heat Mass Tran 2019; 136: 555–562.
57. S.M. Peyghambarzadeh, A. Vatani and M. Jamialahmadi, "Application of asymptotic model for the prediction of fouling rate of calcium sulfate under subcooled flow boiling," Appl. Therm. Eng., vol. 39, pp. 105–113, 2012.
58. Krause S. Fouling an Wlirmeiibertragerflachen durch Kristallisation und Sedimentation. VDI Forschungsheft 1986, 637.
59. Hasson, D., 1981, Precipitation Fouling-A Review, in Fouling of Heat Transfer Equipment, eds. E.F.C. Somerscales, and J. G. Knudsen, Hemisphere, New York, NY, pp. 527–568.
60. R.B. Ritter, "Crystalline fouling studies," Trans. ASME, vol. 105, pp. 374–378, 1983.
61. Hatch GB. Evaluation of scaling tendencies. Mater Prot Perf 1973; 72: 49–55.
62. Yu H, Sheikholeslami R, Doherty WOS. Effect of thermohydraulic conditions on fouling of calcium oxalate and silica. AIChE J 2005; 51(2): 641–648.
63. A.P. Watkinson and O. Martinez, "Scaling of heat exchanger tubes by calcium carbonate," J. Heat Transfer, vol. 97, no. 4, pp. 504–508, 1975.
64. P. Walker and R. Sheikholeslami, "Assessment of the effect of velocity and residence time in $CaSO_4$ precipitating flow reaction," Chem. Eng. Sci., vol. 58, no. 16, pp. 3807–3816, 2003.
65. Hoang TA, Ang M, Rohl AL. Effects of process parameters on gypsum scale formation in pipes. Chem Eng Technol; 34(6): 1003–1009.
66. Bansal B, Chen XD, Müller-Steinhagen H. Analysis of 'classical' deposition rate law for crystallization fouling. Chem Eng Process 2008; 47: 1201–1210.
67. Epstein N. A model of the initial chemical reaction fouling rate for flow within a heated tube, and its verification. In: Proceedings of 10th international heat transfer conference, Brighton, UK, Taylor & Francis, 1994, 225–229.
68. Rose IC, Watkinson AP, Epstein N, Testing a mathematical model for initial chemical reaction fouling using a dilute protein solution. Can J Chem Eng 2000; 78(1): 5–11.
69. Schlichting, H., 1968, Boundary Layer Theory, 6th ed., McGraw-Hill Book Co., Inc., New York, pp. 596–625.
70. Bansal, B., Müller-Steinhagen, H., 1993, Crystallization Fouling in Plate Heat Exchangers, ASME Journal of Heat Transfer, 115, pp. 584–591.
71. Andritsos N, Karabelas AJ, Calcium carbonate scaling in a plate heat exchanger in the presence of particles. Int J Heat Mass Tran 2003; 46: 4613–4627.
72. Li W, Webb RL. Fouling in enhanced tubes using cooling tower water, Part II combined particulate and precipitation fouling. Int J Heat Mass Transf 2000; 43: 3579–3588.
73. Augustin W, Bohnet M. Influence of the ratio of free hydrogen ions on crystallization fouling. Chem Eng Process 1995; 34(2): 79–85.
74. Hasan BO, Nathan GJ, Ashman PJ, Craig RA, Kelso RM. The effect of temperature and hydrodynamics on the crystallization fouling under cross flow conditions. Appl Therm Eng 2012; 36: 210–218.
75. Zhao X, Chen XD. A critical review of basic crystallography to salt crystallization fouling in heat exchangers. Heat Transfer Eng 2013; 34(8–9): 719–732.

76. Awad MM. Fouling of heat transfer surfaces, In: Belmiloudi A, editor. Heat Transfer, theoretical analysis, experimental investigations and industrial systems, France: IntechOpen, 2011, 505–542.
77. Mullin, J. W., Crystallization, 3rd ed., Butterworth-Heinemann, Oxford, UK, 1993.
78. Kazi SN, Duffy GG, Chen XD. Fouling and fouling mitigation on heated metal surfaces. Desalination 2012; 288: 126–134.
79. M.M. Sarafraz, F. Hormozi and S.M. Peyghambarzadeh, "Pool boiling heat transfer to aqueous alumina nano-fluids on the plain and concentric circular micro-structured (CCM) surfaces," Exp. Therm. Fluid Sci., vol. 72, pp. 125–139, 2015.
80. M.M. Sarafraz, S.M. Peyghambarzadeh and S.A. Alavifazel, "Enhancement of nucleate pool boiling heat transfer to dilute binary mixtures using endothermic chemical reactions around the smoothed horizontal cylinder," Heat Mass Transfer, vol. 48, pp. 1755–1765, 2012.
81. M.M. Sarafraz, S.M. Peyghambarzadeh and S.A. Alavifazel, "Enhancement of the pool boiling heat transfer coefficient using the gas injection into the water," Polish J. Chem. Technol., vol. 14, no. 4, pp. 100–109, 2012.
82. M.M. Sarafraz, S.M. Peyghambarzadeh and S.A. Alavifazel, "Experimental studies on nucleate pool boiling heat transfer to ethanol/MEG/DEG ternary mixture as a new," Chem. Ind. Chem. Eng. Q., vol. 18, no. 4, pp. 577–586, 2012.
83. M.M. Sarafraz and S.M. Peyghambarzadeh, "Influence of thermodynamic models on the prediction of pool boiling heat transfer coefficient of dilute binary mixtures," Int. Commun. Heat Mass, vol. 39, pp. 1303–1310, 2012.
84. M.M. Sarafraz, S.M. Peyghambarzadeh, S.A. Alavifazel and N. Vaeli, "Nucleate pool boiling heat transfer of binary nano mixtures under atmospheric pressure around a smooth horizontal cylinder," Period. Polytech. Chem. Eng., vol. 57, no. 1–2, pp. 71–77, 2013.
85. M.M. Sarafraz and S.M. Peyghambarzadeh, "Nucleate pool boiling heat transfer to Al_2O_3-Water and TiO_2-water nanofluids on horizontal smooth tubes with dissimilar homogeneous materials," Chem. Biochem. Eng. Q., vol. 26, no. 3, pp. 199–206, 2012.
86. M.M. Sarafraz, F. Hormozi, S.M. Peyghambarzadeh and E. Salari, "Experimental study on the influence of SO_2 gas injection to pure liquids on pool boiling heat transfer coefficients," Heat Mass Transfer, vol. 50, pp. 747–757, 2014.
87. S.M. Peyghambarzadeh, A. Hatami, A. Ebrahimi and S.A. Alavifazel, "Photographic study of bubble departure diameter in saturated pool boiling to electrolyte solution," Chem. Ind. Chem. Eng. Q., vol. 20, no. 1, pp. 143–153, 2014.
88. Kazi SN, Teng KH, Zakari MS, Sadeghinezhad E, Bakar MA. Study of mineral fouling mitigation on heat exchanger surface. Desalination 2015; 367: 248–254.
89. Zhang GM, Li GQ, Li W, Zhang Z, Leng XL. Particulate fouling and composite fouling assessment in corrugated plate heat exchangers. Int J Heat Mass Tran 2013; 60: 263–273.
90. Esawy, M., 2011, Fouling of Structured Surfaces during Pool Boiling of Aqueous Solutions, Ph.D. thesis, Institute for Thermodynamics and Thermal Engineering, University of Stuttgart.
91. Jamialahmadi M, Müller-Steinhagen H. Scale formation during nucleate boiling-A review. Corros Rev 1993; 11(1–2): 25–54.
92. Höfling V., Augustin, W., Bohnet, M., 2004, Crystallization Fouling of the Aqueous Two-component System $CaSO_4/CaCO_3$, International Journal of Transport Phenomena, 6, pp. 99–109.
93. Bohnet MW. Crystallization fouling on heat transfer surfaces – 25 years research in Braunschweig. In: Müller-Steinhagen H, Malayeri MR, Watkinson AP, editors. Proceedings of 6th international conference on heat exchanger fouling and cleaning - Challenges and opportunities, Kloster Irsee, Germany, 2005, 295–302.
94. Bramson D, Hasson D, Semiat R. The roles of gas bubbling, wall crystallization and particulate deposition in $CaSO_4$ scale formation. Desalination 1995; 100(1–3): 105–113.